Modern Electronics and Integrated Circuits

B J Stanier
Department of Applied Physics and Electronics,
Durham University

Adam Hilger Ltd, Bristol and Boston

British Library Cataloguing in Publication Data
Stanier, B.J.
Modern Electronics and Integrated Circuits.
1. Linear integrated circuits 2. Digital integrated circuits
I. Title
621.381′73 TK7874

ISBN 0-85274-550-8 (hbk)
ISBN 0-85274-552-4 (pbk)

Consultant Editor: Dr C Hilsum, Royal Signals and Radar Establishment, Malvern.

Published by Adam Hilger Ltd
Techno House, Redcliffe Way, Bristol BS1 6NX, England
PO Box 230, Accord, MA 02018, USA

Phototypeset by Quadraset Ltd, Midsomer Norton, Bath
and printed in Great Britain by J W Arrowsmith Ltd, Bristol

Contents

Preface

The exciting developments that have occurred in electronics over the last few years are the result of advances in integrated circuit technology. This book looks at some of the results of that technology and at their impact on analogue and digital electronics. The emphasis is on the application of integrated circuits such as the operational amplifier or the microprocessor, but discrete devices are also considered.

The viewpoint is that electronics is concerned with the processing of information in the form of electrical signals. Very little theory is needed to begin to appreciate the power of electronic processing methods, and analysis of circuits is only introduced where really necessary. Ohm's law, with an extension to reactive components, is essential. However j-notation and complex algebra are not required, although some knowledge of them will help in the understanding of frequency response.

I intend the book to serve as a self-contained text to accompany a course of perhaps 20 to 30 lectures for non-specialists in electronics, or as an introductory survey for those who will be studying the topics in greater depth later. There are no set problems as is common in a text book, most such exercises are somewhat artificial, and I believe it is better to meet and solve problems in the more realistic context of actual design.

I have avoided mention of specific device types as far as possible as the range and properties of electronic devices are constantly changing. There is no substitute for consulting manufacturer's data sheets to discover what devices are actually available and what they can do, nor is there any substitute for actually designing and building electronic circuits and systems oneself.

1

Physics of Semiconductor Devices

From the electrical point of view, most solid materials can be divided into two categories: conductors, which offer very little resistance to the flow of current, and insulators, which present an extremely high resistance. Metals, especially copper and silver, are excellent conductors. Whilst many plastics are such good insulators that it is difficult to detect any current flow at all, good examples being polystyrene and polythene. The conductivity ratio between, say, copper and polystyrene, is greater than 10^{20}.

Conduction in metals takes place because the outer shell electrons in the atoms are only loosely bound to the nuclei. When metal atoms are condensed into a solid with a crystalline structure or otherwise, these valence electrons become detached from their parent nuclei and are then able to move freely throughout the bulk of the material. The cloud of particles is often likened to an electron gas, and because the particles carry charge, it can support conduction when an electric field is applied. In an insulator all of the electrons are tightly bound, and practically none are available to support conduction.

1.1 Energy Bands

Energies in solids are commonly measured in electron volts. One electron volt (eV) is the amount of energy gained by an electron—or any particle with unit charge—when raised through a potential of one volt and is equivalent to approximately 1.6×10^{-19} J.

In an insulator at the absolute zero of temperature, all of the valence electrons' energies lie in a band shown at the bottom of figure 1.1. Some way above this valence band is a higher energy band corresponding to electrons which have been liberated from their parent atoms and are free to move through the solid and contribute to conduction. The energy gap between the valence and conduction bands, known as the band gap, is typically several electron volts wide.

The kinetic energy of the electron gas is given by kT, where k is Boltzmann's constant and T the absolute temperature. At room temperatures kT is about 0.025 eV and is very much smaller than the band gap. The energy distribution of the electrons is similar to that in the kinetic theory of ordinary gases, showing an exponentially falling tail at high energies. Only a tiny proportion of the electrons have enough energy to cross the band gap and the electron density in the conduction band may be as low as $10^6 \mathrm{m}^{-3}$. There is no band gap in a metal. The upper energy bands overlap, and there are large numbers of electrons available for conduction at all temperatures above absolute zero. The electron density will be of the order of the atomic density, that is about $10^{28} \mathrm{m}^{-3}$.

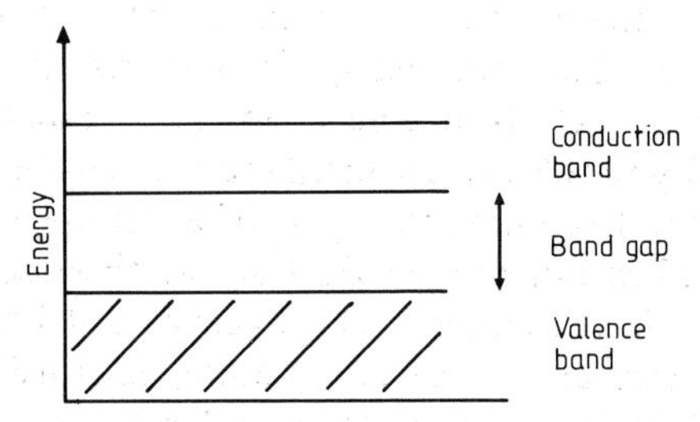

Figure 1.1 Schematic representation of the electron energy levels in an insulator.

In between these extremes some materials, semiconductors, show just a small conductivity. These materials have band gaps of about an electron volt, and at room temperature sufficient electrons can be excited up to the conduction band to give measurable conduction.

Silicon and other semiconductors

Silicon and germanium are semiconducting elements which are important starting materials for electronic devices. Germanium was the first semiconductor to be used in device manufacture but has now been almost completely replaced by silicon. Silicon is preferred, not because of its electronic properties, but for the relative ease with which the fabrication processes (described in the next chapter) can be applied. For some purposes, especially very high speed operation up into the microwave region, semiconducting compounds such as gallium arsenide may also be used. We shall concentrate on silicon and its properties, although the general conclusions about its electronic behaviour also apply to other materials.

1.2 Electrons and Holes

Although we have considered the conduction process to take place through free electrons, this is not the only possible mechanism. When an electron is excited from the valence band to the conduction band it leaves behind a vacancy or hole. This hole corresponds to the ionised parent atom when a valence electron is stripped off. A hole can be filled by another valence electron excited from an adjacent atom, making the hole effectively free to move in the solid. It can thus be treated as a positively charged current carrier.

Every electron excited across the band gap leaves a hole in the valence band and is part of an electron–hole pair. Both the electron in the conduction band and the hole in the valence band are available for conduction purposes. Electron–hole pairs created by thermal excitation do not persist for ever. After a while, a hole and an electron meet and can then recombine. In thermodynamic equilibrium, the rate of generation of electron–hole pairs must be balanced by the rate at which electrons and holes recombine. A constant electron density in the conduction band and an identical hole density in the valence band are then established.

Conduction arising from the generation of electron–hole pairs in a pure semiconductor is called intrinsic conduction but these semiconductors are only of limited use in electronics. The electron and hole densities in an intrinsic semiconductor are given the symbol n_i. Extrinsic semiconductors, where the conductivity is increased by the inclusion of impurity atoms into the solid are the basis of modern semiconductor devices.

1.3 Impurity Doping and Extrinsic Semiconductors

Silicon is tetravalent and the valence electrons form very stable covalent bonds with neighbouring atoms to create a crystal lattice. The stability of these bonds is the reason for the band gap and for the low conductivity at room temperature. A crystal of pure silicon can be doped by replacing a small proportion of the atoms in the lattice by, say, a pentavalent element such as phosphorus. Each pentavalent impurity atom forms four covalent bonds with adjacent silicon atoms but then has a single valence electron spare. Inside the crystal lattice the binding energy of this electron is quite low. At room temperatures the impurity atom is easily ionised so that it can donate the spare electron to the crystal, where it becomes a free electron capable of supporting conduction. Despite the ionisation of the impurity atoms there is no net charge. The ions are still exactly neutralised by the electrons even though they are now mobile.

On an energy level diagram this corresponds to excitation into the conduction band from levels just below the band bottom. The impurity donor levels are shown by broken and dotted lines in figure 1.2 to emphasise that the number of donor atoms and donor levels is small compared with the concentration of silicon atoms, and that they are fixed in the lattice. The gap between the donor levels and the conduction band is sufficiently small that almost all of the donors are ionised at normal temperatures.

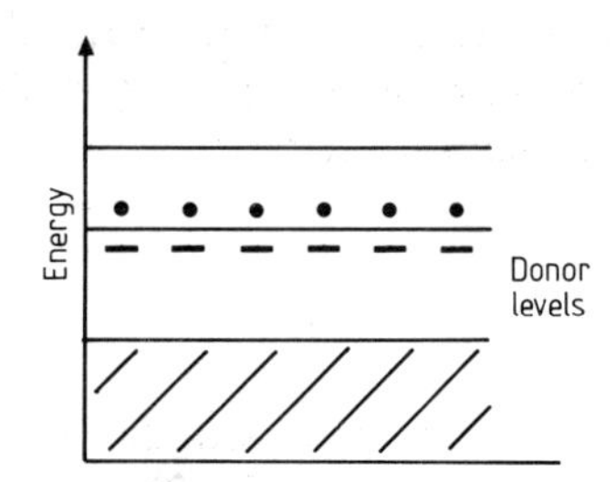

Figure 1.2 Schematic representation of donor levels in an n-type semiconductor (broken line). The donors are ionised, each contributing an electron to the conduction band.

Phosphorus doped silicon is therefore a more conducting material than the pure intrinsic material itself, and the conductivity increases roughly in proportion to the donor impurity concentration. Conduction is predominantly by the donated electrons although there will be a few electron–hole pairs formed by direct excitation across the band gap. Donor doped material is thus described as an n-type semiconductor because the majority of the carriers are negatively charged electrons.

A similar process occurs if a trivalent impurity, boron for example, is introduced into a pure silicon crystal. Now each impurity atom will form covalent bonds with three of the four adjacent silicon atoms, leaving one unpaired electron on the remaining silicon atom. This can bond with a valence electron excited from the silicon lattice leaving a mobile hole. Again these acceptor levels are close to a band edge, this time the top of the valence band. The majority of carriers in acceptor impurity doped silicon are positively charged holes, hence it is p-type material.

Majority and minority carriers

The majority carriers in p-type silicon are holes and those in n-type material, electrons. However in both cases there will be a few carriers of the opposite sign arising from thermal electron–hole pair generation. These minority carriers, in a sea of majority carriers, cannot exist for very long and are annihilated after an average time known as minority carrier lifetime. The higher the majority carrier concentration, the lower the minority carrier concentration that can exist. The relation between the two is

$$n_{\mathrm{p}}n_{\mathrm{n}}=n_i^2 \tag{1.1}$$

where n_{p} and n_{n} are the hole and electron densities and n_i is the intrinsic density defined earlier. Note that n_{p} is the majority carrier density in p-type material and n_{n} the minority carrier density. The reverse is true in an n-type semiconductor where the majority carriers are electrons and the minority carriers are holes. Equation (1.1) shows that the doping level in a semiconductor determines the minority carrier density in addition to controlling the density of majority carriers.

Compensation

A semiconductor which has been doped with donors making it n-type can be converted to p-type by further doping with sufficient acceptors. As acceptors are introduced they bind electrons from the conduction band and immobilise them. When sufficient acceptors have been introduced the donors are compensated and the material becomes intrinsic. Then, adding more acceptors excites electrons from the valence band producing holes and forming p-type material as before. The reverse process is obviously equally possible.

1.4 Diffusion and Drift

There are two mechanisms by which carrier movement, and hence current flow, can occur. Where there is a carrier concentration gradient, carriers will diffuse away from the region of high concentration because of their thermal motion. Diffusion is proportional to the concentration gradient, and for electrons in an n-type semiconductor the diffusion velocity can be written as

$$v=D\,\mathrm{d}n/\mathrm{d}x \tag{1.2}$$

where D is the diffusion constant for the material and $\mathrm{d}n/\mathrm{d}x$ is the concentration gradient. So the diffusion current density will be

$$J=qnD\,\mathrm{d}n/\mathrm{d}x \tag{1.3}$$

where n is the carrier density and q is the electronic charge.

The other mechanism is drift under the influence of an electric field. The charged particles are accelerated by the field but after a while they collide with the crystal lattice and lose the energy gained. The process repeats and the carriers acquire an average drift velocity in the direction of the field. The drift velocity can be written as

$$v = \mu E \tag{1.4}$$

where μ is the mobility and E the electric field. So the drift current density will be

$$J = n\mu E = n\mu \, \mathrm{d}V/\mathrm{d}x. \tag{1.5}$$

1.5 The pn Junction

A pn junction is formed when donor and acceptor impurities are introduced into adjacent regions of a single crystal. It is convenient to imagine that a junction could be formed by placing a region of n-type silicon into intimate contact with p-type silicon. This is not the most practical way of making a junction but it does aid understanding.

An isolated crystal of n-type silicon contains free electrons and one of p-type material has free holes. As the regions make contact there will be a concentration gradient of both electrons and holes across the junction. Holes will diffuse from the p region into the n region where they quickly recombine with electrons. Likewise, electrons in the n region cross the junction into the p region and are soon annihilated.

The loss of holes from the p region and of electrons from the n region leads to a layer on either side of the junction which has been depleted of free carriers. Within the depletion layer there are unneutralised ions which constitute a space charge as shown in figure 1.3. Hence a potential barrier

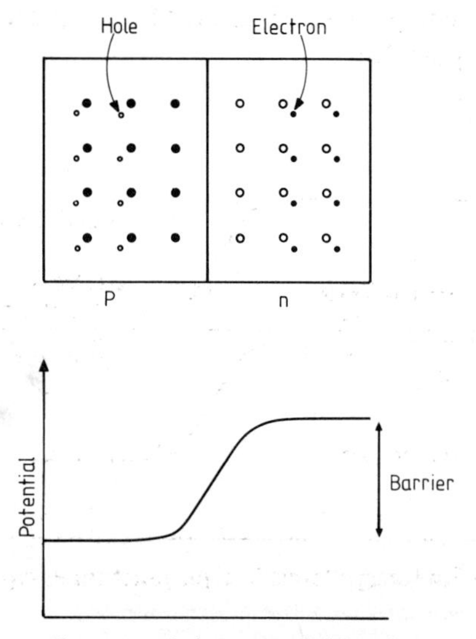

Figure 1.3 Space charge and the potential barrier at a pn junction.

and gradient is now being created across the junction. This potential gradient encourages drift in the opposite direction to the diffusion and eventually thermodynamic equilibrium must be established where the net current flow across the junction and the potential barrier is reduced to zero. The penetration of the depletion layer into the semiconductor depends on the majority carrier density and thus on the doping level. A high doping level requires only a small depletion region to establish equilibrium and vice-versa. Typically the depletion layer is about a μm wide.

In equilibrium a drift current across the junction arises from the electron–hole pairs generated in and close to the depletion layer, and these are then separated and swept away by the electric field. The holes drift to the p region and the electrons to the n region. The electron drift current is exactly balanced by an electron diffusion current in the opposite direction. Similarly, the drift and diffusion hole currents balance and therefore there is no net current flow.

Fermi levels

The effective thermodynamic energy of the free electrons in a material is described by the Fermi level, E_f. For our purposes it is sufficient to say that the Fermi level is about halfway between the source and destination of the thermally excited electrons. So for n-type silicon the Fermi level lies about halfway between the donor levels and the bottom of the conduction band. Similarly, the Fermi level in a p-type material will be just above the top of the valence band. At thermodynamic equilibrium with no externally applied voltage, the Fermi levels of the p-type region must be equal with those of the n-type region. The energy level diagram of a pn junction in equilibrium is then as shown in figure 1.4.

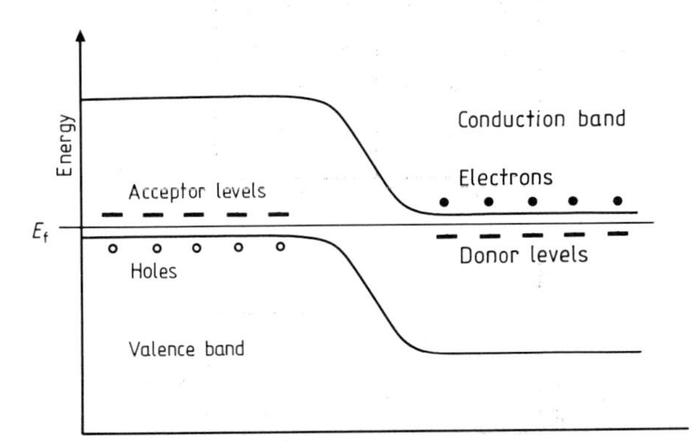

Figure 1.4 Energy levels in a pn junction at equilibrium.

2

Fabrication and Properties of Semiconductor Devices

The properties of semiconductors have been exploited to make components and devices which form the basis of modern electronics. Initially devices were made from germanium, but the fabrication process was limited. Consequently the planar process for silicon described in §2.1 was a major step forward in the way transistors could be fabricated, and the process has been extended to produce other circuit components.

At first the planar process was applied to the production of single (discrete) transistors and similar devices. Later, as the technology was improved, it became possible to make about a hundred devices on the same chip of silicon and the first small scale integrated (SSI) circuits were produced. Further refinements allowed higher packing densities to be achieved and led to medium scale integration (MSI), large scale integration (LSI), and now very large scale integration (VLSI). Each level of integration represents a packing density increase by a factor of ten, so VLSI chips may contain upwards of a hundred thousand devices.

All of these levels of integration share the same basic fabrication method. The improvements in packing density and hence in circuit complexity, have come from technology improvements in each of the fabrication stages outlined below.

2.1 Integrated Circuit Technology

An essential first stage in the production of an integrated circuit (IC) is the manufacture of large and very pure single crystals of silicon. Single crystals are needed because any imperfections in the crystal structure seriously degrade the electronic properties as described in Chapter 1. Large crystals are needed because the cost of production of a single integrated circuit is reduced if many identical circuits can be fabricated simultaneously.

Crystal growth

Large crystals are grown by melting in a crucible a charge of highly refined silicon with a suitable acceptor impurity. A small seed crystal is then dipped into the melt and slowly withdrawn, forming a cylindrical crystal which may be up to 100 mm in diameter. After removal of the low quality ends of the cylinder, the crystal is sawn into thin blanks which are then lapped and polished to produce wafers. These wafers serve as substrates onto which the integrated circuits will be formed.

Epitaxial growth

A thin layer of silicon is then grown on one prepared surface of a wafer by reaction and deposition from a gaseous silicon compound. This epitaxial layer is a continuation of the crystal structure of the underlying substrate wafer and is made n-type by incorporating a donor impurity at the time of growth. It is within this epitaxial layer that the devices and components of the integrated circuit are fabricated.

Oxide masking and diffusion

A thin layer of silicon oxide is grown on the wafer surface by reaction with water vapour or oxygen at high temperature. Photolithographic techniques are then used to etch a pattern of windows into the oxide. A thin coating of photoresist on the oxide surface is exposed to ultraviolet light through a patterned mask. The unexposed portions of the resist can then be dissolved and the uncovered oxide etched away. After removal of the remaining photoresist, the structure is as shown in figure 2.1.

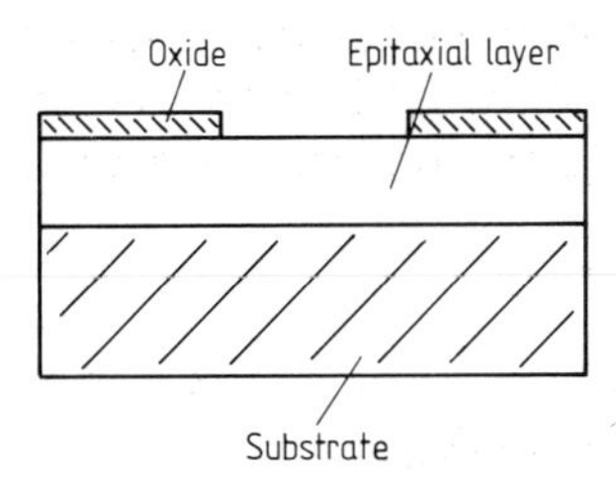

Figure 2.1 Oxide masking on a silicon n-type epitaxial layer which has been grown on a p-type substrate.

Suitable impurity atoms, acceptors or donors, can then be introduced into the silicon by a diffusion process. This diffusion should not be confused with the diffusion of charged carriers mentioned in §1.4. Here we are concerned with the diffusion of atoms through a crystal lattice at high temperature. The wafer is exposed in a furnace to a vapour containing the required impurity material.

Impurity atoms then migrate down into the epitaxial layer from the exposed surface where the oxide mask has been removed. The impurity concentration and the depth of its penetration are determined by the temperature and the diffusion time. Figure 2.2 shows an example where a pn junction has been formed by the diffusion of a p-type region into the n-type epitaxial layer.

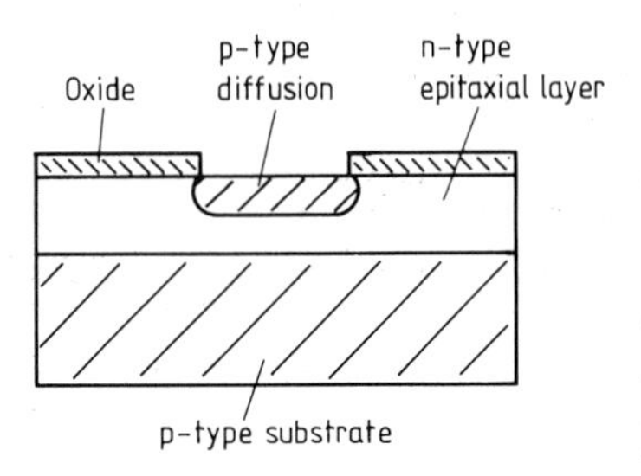

Figure 2.2 A pn junction formed by diffusion through a window in the oxide mark.

The complete process of oxide growth, resist coating and exposure, etching and diffusion can be repeated to form a multilayer structure. Electrical contacts can be made to the surface layers by a further oxide masking stage, with aluminium evaporated through the windows onto the exposed silicon. The same aluminium evaporation can be etched to form an interconnection pattern between devices.

2.2 The pn Junction

In thermal equilibrium, and with no externally applied voltage between the two regions, there is no net current flow through a pn junction. As was shown in §1.5, there is a small diffusion of majority carriers from each region across the junction because a few carriers have sufficient thermal energy to surmount the potential barrier at the depletion layer. This current is balanced by an opposite and equal flow from the thermal generation of electron–hole pairs in and near the depletion layer, which are then separated and swept away from the junction by the electric field.

An external applied voltage will change the height of the potential barrier. If the p region is made negative with respect to the n region, the barrier height is increased as shown in figure 2.3. This must reduce both the hole flow from left to right and the electron flow from right to left. But the drift currents resulting from electron–hole generation will be unaltered. So now there is a net flow of conventional current from the n region to the p region. Because the tail of the energy distribution of the majority carriers falls exponentially as kT, an increase in the barrier height by only a few times kT is sufficient to reduce the diffusion current to a negligible value. Then the net current saturates at a value determined by the rate of electron–hole generation.

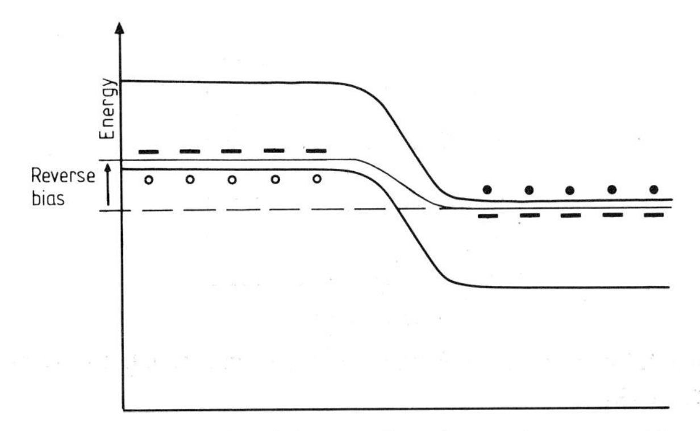

Figure 2.3 Energy levels in a pn junction under reverse bias.

The junction is then said to be reverse biased and is almost non-conducting. The small leakage current that does flow is called the saturation current and is of the order of nanoamps for most silicon pn junctions at room temperature. For many purposes this leakage current can be neglected, but it does increase rapidly with temperature and may become important if the junction becomes hot.

If the polarity of the applied voltage is reversed, the potential barrier will be reduced. Now more majority carriers will have sufficient energy to cross the barrier. Again though, the electron–hole generation process remains at the equilibrium value and a net current flows. As the applied voltage is raised, this current increases approximately exponentially. The junction is now forward biased.

A detailed consideration of the statistics of the carrier energy distributions in a pn junction leads to an expression for the current I valid for both forward and reverse applied bias voltage V

$$I = I_s[\exp(qV/\eta kT) - 1] \qquad (2.1)$$

where I_s is the saturation leakage current and η is a factor of about 2 for silicon devices.

Figure 2.4(a) shows the form of this relation. The scales have been chosen so that I_s is obvious. We are not usually interested in such small currents and figure 2.4(*b*) shows the same relation on a more normal scale. I_s now appears to be negligible, and there is little current flow apparent in the forward direction until the applied voltage reaches about 0.5–0.6 V whereafter the current rises rapidly. This relation is highly non-linear and although continuous, as emphasised by figure 2.4(*a*), it is often convenient to regard the characteristic curve as showing two distinct regions. For all applied voltages below about 0.5 V, the device has no significant conduction and behaves as a very high resistance, whereas above that point it conducts with a low resistance.

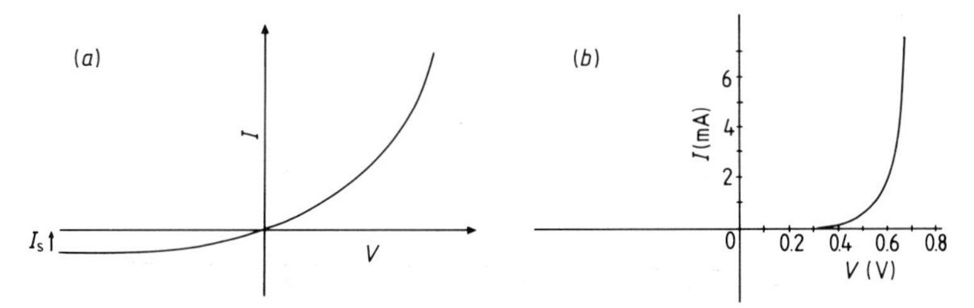

Figure 2.4 Current–voltage relation for a pn junction (*a*) with expanded scales, (*b*) with normal scales.

The pn junction diode

This two-valued resistance is very useful and is the major property of the pn junction diode. The diode has two connections, one to the p region and the other to the n region. The standard diode symbol shown in figure 2.5 has an arrowhead to indicate the direction of conventional current flow when the device is biased in the forward direction.

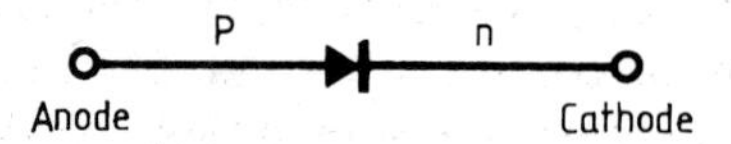

Figure 2.5 The arrowhead in the diode symbol indicates the direction of forward current flow from the anode to the cathode.

If the reverse voltage on a pn junction diode is increased sufficiently, the carriers produced by electron–hole pair production in and near the depletion layer will acquire enough energy from the electric field to cause ionisation and hence generate further carriers. The leakage current then begins to increase, and the process may avalanche, leading to a very rapid increase in current.

Breakdown can occur at quite low reverse voltages if the doping levels are high. Then the width of the depletion region becomes very thin, and carriers can penetrate the potential barrier by quantum mechanical tunnelling. This is the Zener effect.

It is possible to produce diodes which have precisely controlled breakdowns from below 3 V to above 150 V. Such diodes are useful as voltage references. Diodes with low reference voltages use the Zener effect. Those with values above about 5 V exploit avalanching, but all are normally called Zener diodes. A typical current–voltage characteristic for a Zener diode is shown in figure 2.6.

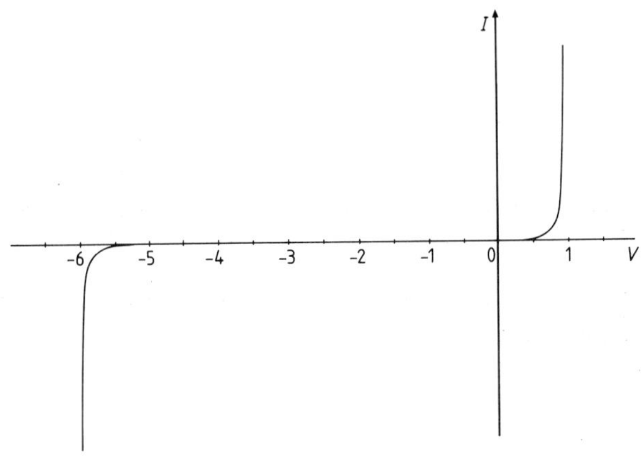

Figure 2.6 Current–voltage relation for a typical Zener diode with a nominal breakdown of 6.2 V.

2.3 The Bipolar Transistor

The bipolar transistor (BPT) is a three-layer structure comprising two pn junctions with a central control region sandwiched between two outer regions. The regions may be p-type or n-type, so there are two possible arrangements, one with an npn sandwich, the other a pnp version. Most of our discussion of BPT properties will concentrate on the npn type, but the general conclusions are equally applicable to a pnp device except that the polarities of all voltages and currents must be reversed.

The two-layer structure of figure 2.2 can easily be extended to make a BPT by adding a further n-type diffusion as shown in figure 2.7. The vertical scale in the figure has been expanded for clarity. In practice the vertical dimension is very small, and the p-type region in the middle of the sandwich is extremely thin. Connections are made to all three regions. The first n-type layer is the collector, the middle p-type region is the base and the final n-diffusion is the emitter. Both the base–emitter and the base–collector junctions exhibit

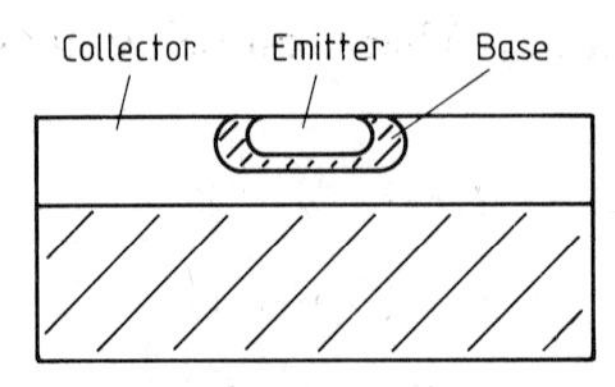

Figure 2.7 An npn transistor formed by a p-type diffusion and an n-type diffusion into an n-type epitaxial layer, the p-type regions are represented by the hatched areas.

normal pn junction diode characteristics when operated in isolation, with the other electrode in open-circuit.

Suppose now that the base–collector junction is reverse biased by making the collector positive with respect to the base. Only a very small leakage current flows and the collector current is negligible. If now the base–emitter junction is forward biased, it conducts and emitter and base currents flow. But the two junctions interact through the very narrow base region and current flows in the reverse-biased collector junction too.

This remarkable effect can be explained in terms of the energy level diagram shown in figure 2.8. The base–emitter junction is forward biased at V_{be}, so majority carriers (electrons) from the emitter are injected into the base. If the width of the base region is sufficiently small, a large fraction of them can cross the base region and arrive at the base–collector junction without recombining with majority holes. At the collector–base junction they experience a strong electric field and are swept into the collector region to form the collector current.

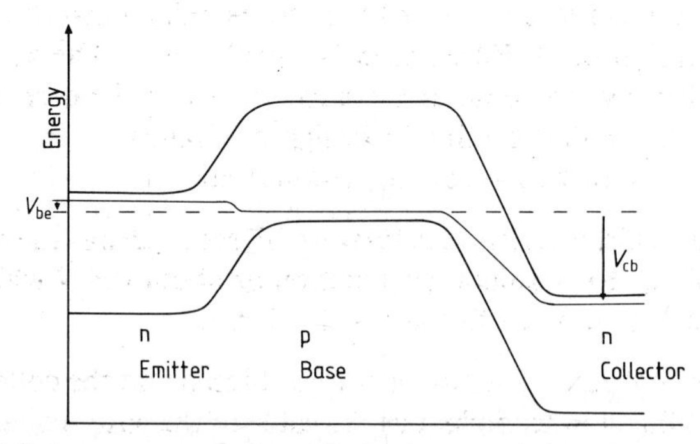

Figure 2.8 An npn transistor with the base–emitter junction forward biased and the base–collector junction reverse biased.

The proportion of the emitter current I_e which reaches the collector is given the symbol α. So the collector current $I_c = \alpha I_e$. The fraction of the emitter current which fails to reach the collector because of recombination appears as the base current I_b. There can be no net charge accumulation in the transistor, so

$$\begin{aligned} I_b &= I_e - I_c \\ &= (I_c/\alpha) - I_c \\ &= (\alpha^{-1} - 1)I_c \\ &= \beta I_c \end{aligned} \tag{2.2}$$

where β is commonly called the current gain of the transistor. Sometimes the symbols α' and h_{FE} are used for this quantity.

The proportion of the carriers which recombine after entering the base region from the emitter depends on the relation between the minority carrier lifetime within the material and the time which they take to cross the base region. These quantities are mainly determined by the material properties and the geometry of the device. So the current gain β is fairly constant over a wide range of values of V_{be}, I_b, V_{cb} and I_c.

The collector current is therefore primarily determined by the forward bias V_{be} on the base–emitter junction and hence also by the base current I_b. It is hardly affected by the reverse bias on the base–collector junction although there is a small increase in the collector current as the collector–base voltage is increased.

Sometimes it is convenient to consider the collector current as being determined by the value of V_{be}. This gives a voltage control model for the behaviour of the BPT. However, the relation between I_b and V_{be} follows the same exponential form as the law for a pn junction diode. I_c is directly proportional to I_b and so will vary approximately exponentially with V_{be}. It is often more appropriate to regard the BPT as being current controlled as implied by equation (2.2). Either view is equally valid. The choice between them is decided by personal preference or by whichever is the most convenient for the understanding or design of circuits.

These properties of the BPT can be summarised.

(i) The base–emitter junction is forward biased. It drops the normal value of forward voltage for a silicon pn junction of about 0.6 V which is almost independent of the value of the forward current.

(ii) The base–collector junction is reverse biased and the collector current which flows is almost independent of the value of the base–collector voltage.

(iii) The collector current is an exponential function of the forward bias on the base–emitter junction and is directly proportional to the base current.

These properties are illustrated in figure 2.9 which shows the relation between I_b and V_{be} (the input characteristic), and that between I_c and V_{ce} (the output characteristic), for a number of values of I_b. Note that it is conventional to refer all the voltages to that of the emitter, so figure 2.9(*b*) uses V_{ce} rather than V_{cb} as the voltage scale.

If V_{ce} is increased sufficiently, the collector–base junction will reach breakdown and the collector current ceases to be properly controlled by the base–emitter bias. Normal operation of a BPT is confined to voltages below this breakdown region.

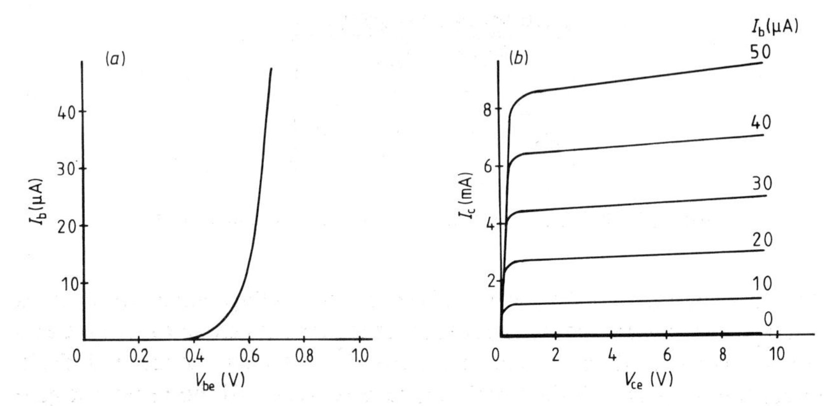

Figure 2.9 Typical current–voltage relations for (*a*) the input (base–emitter), (*b*) the output (collector–emitter) for a small npn transistor.

Discrete bipolar transistors are widely used for amplification and switching purposes and are also the standard devices for many integrated circuits. For some applications though, the input current drawn by the base is undesirable.

Also, the BPT needs to be electrically isolated from other devices and components fabricated on the same chip. This isolation can be produced by a deep p diffusion down through the epitaxial layer to the substrate around each transistor. The pn junction formed is reverse biased in normal operation of the device and provides the required isolation. It occupies a considerable area on the silicon chip. For applications which require very low input currents, or for integrated circuits where very small area devices are needed to give high packing densities, an alternative transistor structure can be used.

2.4 The Metal Oxide Semiconductor Transistor

The metal oxide semiconductor transistor (MOST) is made by the same oxide masking and impurity diffusion process as the BPT. Figure 2.10 shows a cross section of a MOST (again the vertical scale is exaggerated for clarity). The essential feature is two n-type diffusions into a lightly doped p-type substrate. The n regions are very close together and the p region between them is covered by a thin layer of insulating silicon oxide and a conducting layer of metal. Connections are made to the two n regions (source and drain) and to the metal (gate).

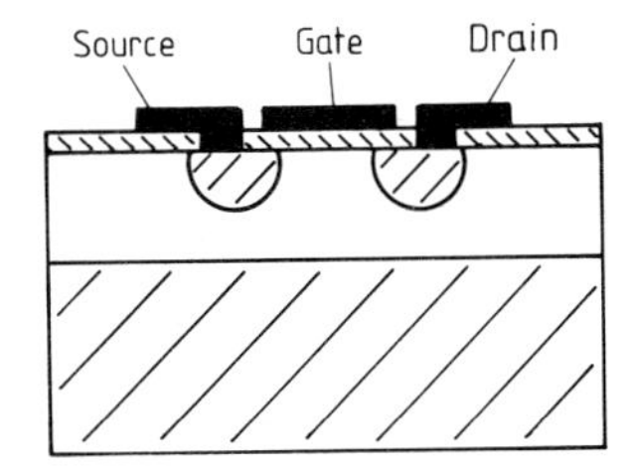

Figure 2.10 Schematic cross section of an n channel MOST.

In many applications the substrate is connected to the source. The drain is made positive with respect to the source. If the gate is held at the source and substrate voltage, then no current can flow between source and drain because the substrate–drain junction is reverse biased. If, however, the gate is made positive with respect to the source and substrate, the electric field under the gate repels the majority carriers (holes) in the surface of the substrate and creates an n-type channel between source and drain. Current can now flow between source and drain in this induced channel.

For values of the drain–source voltage V_{ds} well below the gate–source voltage V_{gs} the channel behaves like a simple resistor with the drain current I_d increasing approximately linearly with V_{ds}. However, an increase in V_{ds} reduces the gate–drain voltage V_{gd} and the field and channel depth near the drain are therefore reduced. When V_{ds} approaches V_{gs} the channel is constricted and the drain current becomes limited. Higher values of V_{ds} will then only produce a small increase in I_d.

This type of n channel MOST is operated with a positive value of V_{gs} to enhance the conduction. Figure 2.11 shows the output characteristics of an enhancement n channel MOST. A modification of the fabrication process can produce depletion devices where current flows for $V_{gs}=0$. Depletion MOST are operated with negative values of V_{gs} reducing the values of I_d. As with the BPT, a reversed polarity p channel MOST with p-type drain and source regions diffused into an n-type substrate is also possible.

The input characteristic of all types of MOST is just that of a capacitor. Only an extremely small leakage current flows through the insulating oxide. However, the very thin oxide layer is easily destroyed by excessive voltages, such as may be caused by static electricity. Although device manufacturers incorporate diodes in the basic structure of figure 2.10 to give protection, care must be taken when handling integrated circuits using MOST.

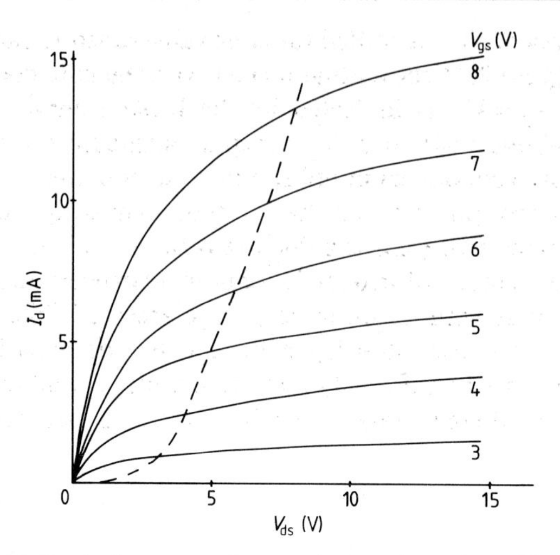

Figure 2.11 Typical output characteristics, I_d against V_{ds}, for an n channel enhancement MOST. The broken curve is the locus of $V_{gs} = V_{ds}$ (see §2.5).

The junction field effect transistor (JFET) is another transistor which relies on the modulation of a conducting channel. A basic JFET structure is similar in cross section to the pn junction shown in figure 2.2. A gate connection is made to the p region and drain and source connections are made to the n region on each side. The gate is maintained negative with respect to the source so that the pn junction is reverse biased. The gate region is much more heavily doped than the n layer, so the depletion layer extends well into the n layer. For small values of the reverse bias, current can flow between source and drain in the semiconducting n layer under the p region. As the reverse bias is increased, the depletion layer extends further into the n layer increasing its resistance and reducing the current. As with the MOST, the JFET exhibits saturation of the drain current as the drain to source voltage is raised because the reverse bias on the junction near the drain increases and the channel becomes constricted. It is obvious that the substrate must be maintained at or below the potential of the source to ensure that the current flow is restricted to the n layer.

The input current to a JFET is just the reverse leakage of the pn junction and is low. However, it is not as low as the gate current in a MOST.

2.5 Resistors and Capacitors

Resistors can be fabricated by the oxide masking and diffusion process. All that is required is a semiconducting track with connections at either end.

However, the area of silicon needed for large value resistors can be excessive. Inspection of figure 2.11 shows that a MOST with the gate connected to the drain, so that $V_{gs} = V_{ds}$ as indicated by the broken curve, behaves as a somewhat non-linear resistance. This is quite adequate for many circuits, and is much more economical in utilisation of silicon area.

A reverse biased pn junction has a thin insulating depletion layer sandwiched between two (semi)conducting regions. This is equivalent to a parallel plate capacitor. Unfortunately, only small capacitors of up to about 100 pF can be made without using very large areas of silicon. Also, such capacitors are non-linear, showing a dependence of capacitance on the applied voltage. This non-linearity may be a nuisance, although there are some applications where voltage controlled variable capacitance diodes can be useful.

3

Signals in Electronic Systems

Electronics is concerned with the processing of information in the form of electrical signals. The information is usually to do with non-electrical quantities such as linear and angular displacement, pressure, temperature and velocity. These quantities cannot be directly processed by electronic devices. Transducers are needed to convert between these quantities and electrical signals. Examples of simple input transducers include a mechanically operated switch which converts position, a rotary potentiometer or variable resistance which produces a signal dependent on angular displacement and a thermocouple which responds to temperature difference. Loudspeakers, motors and VDU are output transducers which convert electrical signals into sound, rotary movement and light respectively.

A general electronic system as shown in figure 3.1 accepts signals obtained from input transducers, processes them, and returns them through output transducers. A system may be further divided into subsystems and then into circuits. The boundaries between these classifications are not sharply defined, but a circuit is generally understood to contain just a few active devices and their associated components. A subsystem may contain several

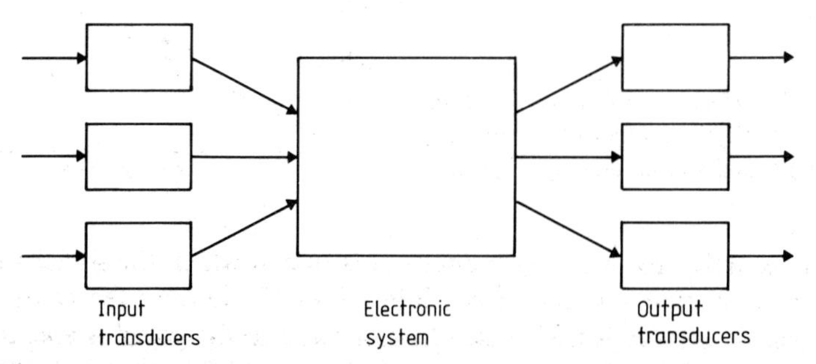

Figure 3.1 A general electronic system accepts input signals obtained from input transducers, processes them, and returns them through output transducers.

circuits and performs an identifiable signal processing operation. It is important to realise that signals must vary with time to contain any useful information. The variation may be slow, as, for example, in the output from a thermocouple measuring the temperature of a refrigerator, or it may be very rapid as in the case of the signal representing picture information in the output from a television camera.

Steady state values of voltage or current in a system or circuit are not signals in that they do not carry information. Such constant levels are sometimes called DC (direct current) to distinguish them from time varying signals known as AC (alternating current). Wherever possible we shall use lower case symbols in expressions and equations to indicate signals, and capitals to signify steady state quantities.

3.1 Analogue and Digital Signals

Most signals which represent physical quantities vary continuously, so there is an infinite number of possible signal levels (most physical quantities are essentially continuous on the macroscopic scale). Such signals are usually described as analogue, implying that the signal is an analogue of the physical quantity it represents, and that there is a one to one correspondence between any value of the quantity and the signal representing it. The relation between the signal s and the quantity q may be linear as shown in figure 3.2, but this is not essential.

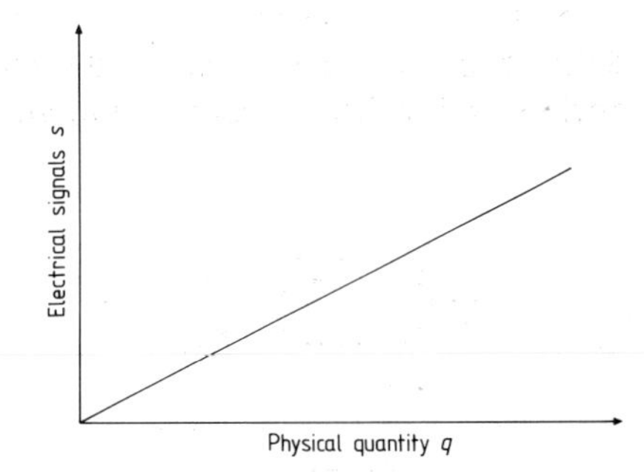

Figure 3.2 The continuous relation between a physical quantity q and an electrical signal s in an analogue system.

An alternative form of representation is one in which the signal value is restricted to a finite number of possible values. There are a few cases where the physical quantity is itself discontinuous and so the signal is able to take only discrete values. There are some transducers which convert a continuous quantity directly into a discrete signal. But usually discrete signals are obtained from continuous signals through analogue to digital conversion.

Digital to analogue and analogue to digital converters are examined further in Chapter 9.

Figure 3.3 shows an eight-level representation of an analogue quantity. The signal s is constrained to take only the values 0 to 7 inclusive, so that, for example, all values of q in the range $15 \leq q < 20$ are represented by the same signal $s = 3$. Clearly, in this eight-level arrangement the accuracy of the representation may be as poor as one part in eight. Practical discrete signals may need many thousands of levels to provide sufficient precision.

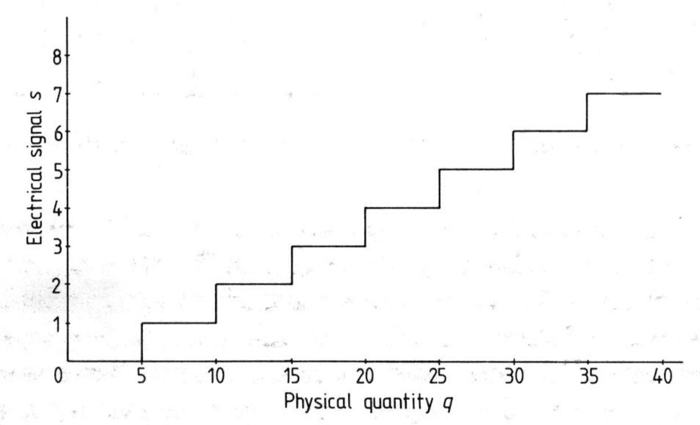

Figure 3.3 An eight-level representation of a quantity in a discrete system.

Although the signal in a discrete system can usually take any one of a large (but finite) number of levels, it is encoded as discussed in Chapter 7 so that only two distinct voltage levels are needed for its representation. This is the basis of digital electronics where signals are processed in the form of binary (two-state) codes.

3.2 Periodic and Aperiodic Signals

Many real signals show a periodic pattern in that they repeat, exactly or nearly exactly, at regular intervals. The shape of the signal over one period is the waveform. Figure 3.4(a) shows an example of a waveform such as might be found in a segment of human speech, a continuous (analogue) signal. Figure 3.4(b) shows an example of a digital signal which also exhibits a periodic waveform.

Perfectly periodic waveforms are convenient test signals, and are generated by many laboratory instruments. They are useful because the same test signal can be reproduced as often as desired. Also, events taking place at the high speeds common in electronics can be repeated again and again, and the events observed, say, on an oscilloscope, where the scan time of the beam can be synchronised to an integral multiple of the period of the waveform. The repetitive events then overlap and merge to give an apparently stationary display effectively freezing time.

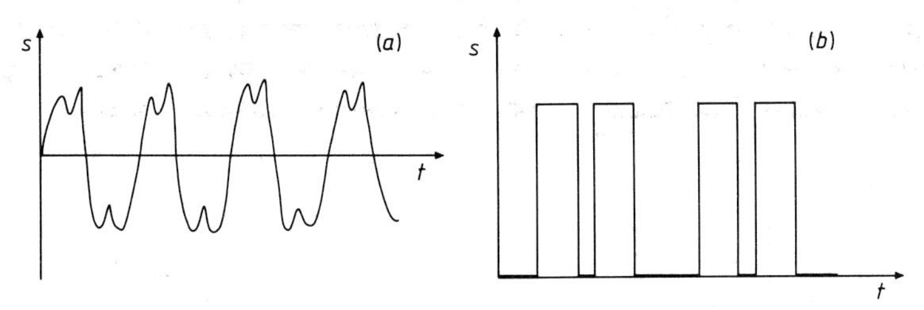

Figure 3.4(*a*) A periodic analogue signal; (*b*) a periodic digital signal.

A few signals vary in an apparently random fashion with time and show no appreciable periodicity. Such signals are difficult to observe, particularly if the variation is rapid. Use is sometimes made of test signals which appear to be random over an interval of time, but which nevertheless repeat exactly over longer intervals. Such pseudo-random signals have some of the properties of truly random signals, but their periodicity makes them reproducible and more easily observable.

The majority of real signals lie between the extremes of perfect periodicity and pure randomness. Speech is a good example in that waveforms may be identified, but are continually changing. There is short term periodicity, but the period and the waveform alter in the longer term.

3.3 Sinusoidal Signals

A sinusoidal waveform of the form shown in figure 3.5 is a particularly simple periodic signal which is important in electronics. It is defined by the equation

$$s = s_p \sin \theta = s_p \sin \omega t \tag{3.1}$$

where s_p is the peak excursion of the signal and the angle θ is called the phase. If the period is T seconds then the rate at which the waveform repeats is the natural frequency $f = 1/T$ Hz. It is common in electronics to work with the angular frequency ω rather than with natural frequency and therefore one period T corresponds to a 360° or 2π change in θ and ωt. So the angular frequency $\omega = 2\pi f$ radians/second.

A very wide range of frequencies are met in electronics, we may be dealing with signals with periods measured in minutes or hours and thus can be regarded as having essentially zero frequency (again sometimes confusingly called DC). At the other extreme, some electronic communications systems have frequencies measured in GHz (10^9 Hz).

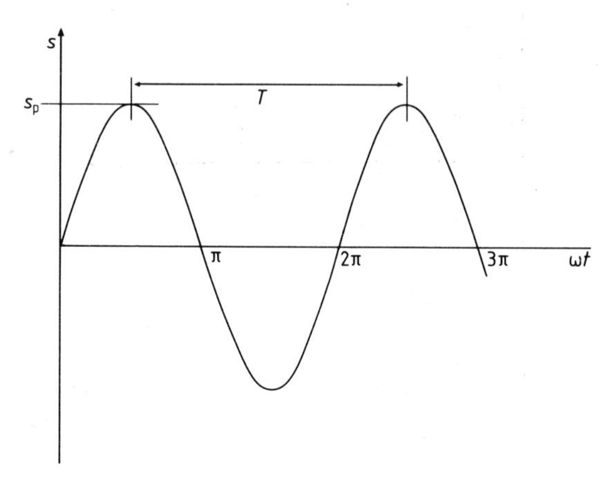

Figure 3.5 The sinusoidal waveform $s = s_p \sin \omega t$.

The sinusoid is interesting because it leads to a simple description of the behaviour of capacitance and inductance when driven by sinusoidal signals. Then the relation between the magnitudes of an applied sinusoidal voltage v and the corresponding sinusoidal current i can be written in a form very like Ohm's law as applied to steady state voltage, current and resistance.

$$v = iZ \tag{3.2}$$

where Z is the reactance. For a capacitance C the magnitude of the reactance is $1/\omega C$, and for an inductance L it is ωL. Sinusoids are widely used as test signals for investigating the behaviour of electronic systems and circuits. Near perfect sinusoidal waveforms over a very wide range of frequencies are available from standard laboratory signal sources (oscillators).

3.4 Phase and Phase Difference

Absolute phase is not usually of much interest. We are more concerned with the difference in phase or phase shift between corresponding points on two waveforms of identical frequency (or the same signal at different points in a circuit). Figure 3.6 shows two sinusoids of identical frequency which have a relative time displacement. The difference between their phase angles, the phase shift ϕ, is 90° or $\pi/2$ in this case. Waveforms are said to be in phase if the phase shift is 0° and out of phase when the phase shift is 180° or π. It should be noticed that for a symmetrical periodic signal such as a sinusoid, a phase shift of π is equivalent to inversion or sign reversal.

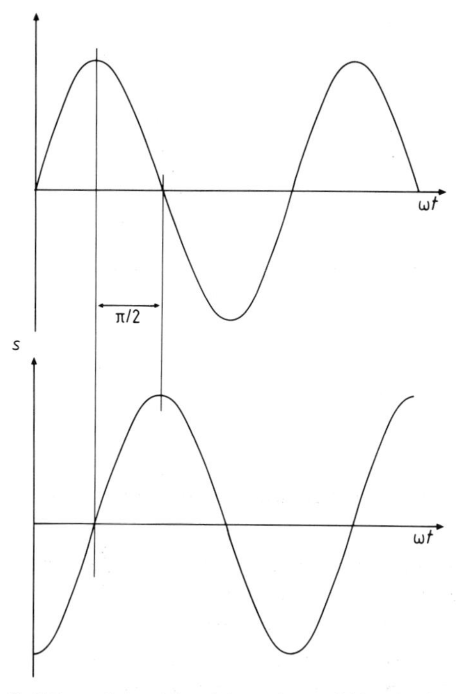

Figure 3.6 Two sinusoids with a phase shift of $\pi/2$ or 90°.

3.5 Pulse Waveforms

A pulse is a rapid—ideally instantaneous—transition from one signal level to another followed after an interval by another transition back to the original level. In general the transitions may be between any arbitrary levels, but the most common case is when the base level is zero as shown in figure 3.7. The time t spent at the other level is the pulse length. Practical pulses cannot have instantaneous transitions between their levels because it is impossible to produce a discontinuous change in the voltage at a point or in the current flowing in a conductor. Real signals are always continuous and all real pulses will show smooth transitions between the levels if examined on a sufficiently fine time scale as illustrated in figure 3.8.

The speed of the transitions between the levels is characterised by the rise and fall times, t_r and t_f respectively. The beginning and end of a smooth transition are difficult to identify accurately, so the times are conventionally taken between the 10% and 90% fractions of the pulse height.

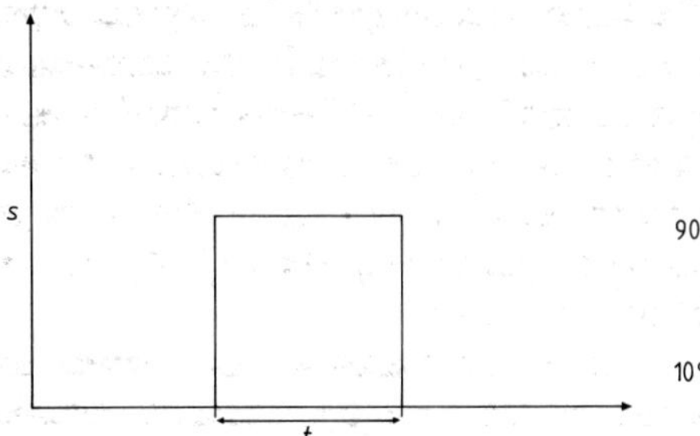

Figure 3.7 An ideal pulse of length t.

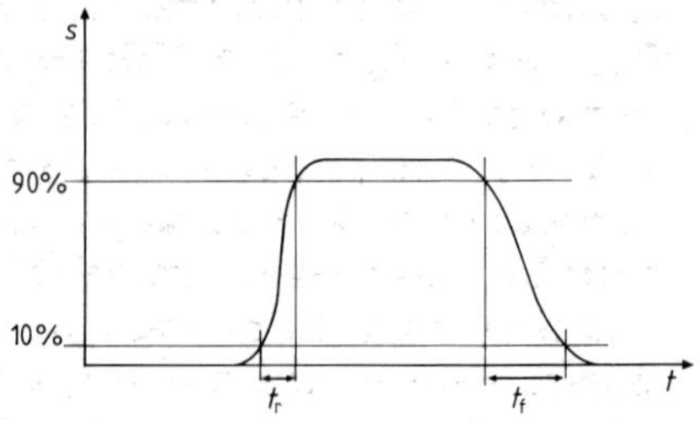

Figure 3.8 A real pulse with rise and fall times t_r and t_f.

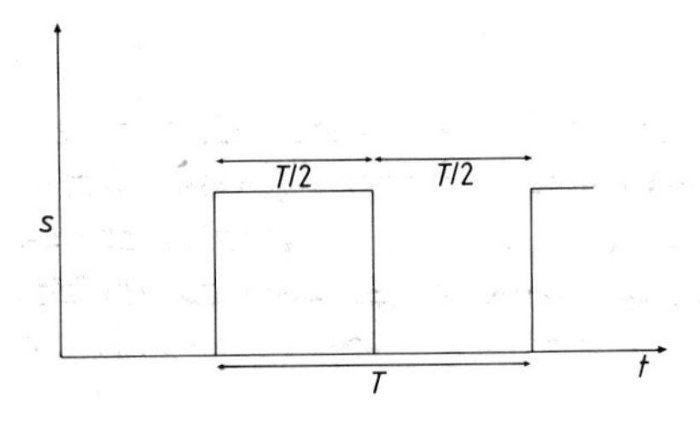

Figure 3.9 A square wave of period T.

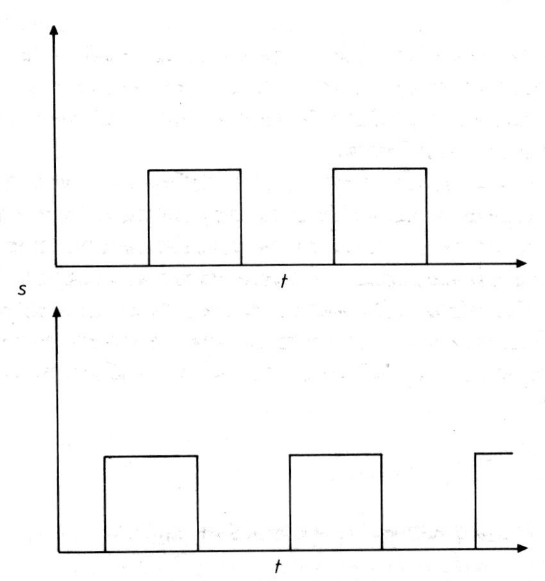

Figure 3.10 Two square waves with a phase shift of a quarter of a period ($\pi/2$).

The concepts of waveform and frequency can be applied to regular trains of pulses. One of the most common pulse waveforms is the squarewave illustrated in figure 3.9. This has a period T and a fundamental natural frequency $f = 1/T$. Strictly, phase and phase shift are only properly defined with respect to sinusoids, but the terms are often loosely applied to pulse waveforms too. So, for example, the two squarewaves shown in figure 3.10 might be said to have a phase shift of $90°$ or $\pi/2$.

Regular pulse trains and squarewaves are essential test signals for examining the behaviour of digital circuits. They are also useful as periodic test signals for analogue circuits because they contain a wide range of frequency components as discussed in §3.7.

3.6 Peak and RMS Signal Amplitude

It is customary to think of signals in terms of their voltage levels and most electronic instruments used for the measurement and observation of signals are scaled in voltage. Sometimes it is more convenient to work with a current representation of signal level. Occasionally it may be necessary to consider the power level of a signal. We shall use voltage signal levels almost exclusively in subsequent discussion.

The amplitude of a signal may be expressed in several ways. The peak to peak excursion is convenient for periodic signals observed on an oscilloscope because it can be directly measured from the trace. In most other cases it is better to use the steady state (DC) value which carries the same power. Power is proportional to the square of voltage or current and so this equivalent value can be obtained by taking the square root of the mean of the square of a signal over some time interval.

This root mean square (RMS) value is widely used in scaling measuring instruments. However, the calibration may not be valid for all waveforms. Many instruments are scaled in RMS but actually measure average or peak to peak values. Their calibration is based on the assumption that the signal waveform is sinusoidal. (The RMS value for a sinusoid is easily shown to be $1/\sqrt{2}$ of the peak value.) Instrument manufacturers sometimes use the description 'true RMS' to indicate a calibration valid for other waveforms as well.

3.7 Fourier Components of Non-sinusoidal Signals

So far we have considered a signal in terms of its variation with time. An alternative and exactly equivalent view of the properties of a signal can be obtained by considering its frequency distribution or spectrum. An ideal sinusoidal waveform has a single unique frequency defined as above by $f = 1/T$ and its spectrum consists of a single line. All other waveforms are equivalent to a summation or superposition of a number of sinusoids, so they contain a corresponding number of frequency components (another reason for the theoretical importance of the sinusoid).

A periodic waveform can be shown to contain a set of frequency components which are integral multiples of the fundamental frequency. In addition to the fundamental component of frequency f there will be $2f$, $3f$, $4f$ and so on. These multiples of the fundamental frequency f are called harmonics and produce a set of lines in the signal spectrum. Aperiodic signals give rise to a continuum of components rather than the discrete regularly spaced lines that come from periodic signals.

It is possible to calculate the magnitudes and relative phases of all of the components in a particular signal. In the periodic case, the fundamental and its set of harmonics are defined by a Fourier series. A Fourier integral is required for an aperiodic signal. It is sufficient here to state some general conclusions which are intuitively obvious. Rapid rates of signal change such as the leading and trailing edges of a pulse are associated with high frequency components. Very slow time variations imply the presence of low frequencies. If a signal has a non-zero average value, there will be a spectral component at zero frequency (DC).

3.8 Bandwidth

The difference between the highest and lowest frequency components in the spectrum of a signal is the bandwidth which is, roughly, a measure of the information content of a signal. A nearly perfect sinusoidal signal from a laboratory oscillator might occupy a narrow band only a few Hz wide and conveys very little information. Speech carries rather more information and has a bandwidth of a few kHz. A picture (video) signal from a television camera has a bandwidth of several MHz and contains a large amount of information.

Bandwidth is important because it is a major requirement to be considered in the design of a circuit or system. A wideband signal will need equally wideband circuits capable of accepting the range of frequency components. If the bandwidth of the circuits is restricted, some of the components in the signal will be reduced or removed and the information content of the signal diminished. A pulse waveform presents a good illustration of the effects of bandwidth restriction.

If a pulse is processed by a circuit which cannot respond to its high frequency components, the rise and fall times are lengthened. This corresponds to a reduction in the precision with which the timing of the leading and trailing edges of the pulse can be determined. A loss of low frequency components introduces curvature in the horizontal portions of a pulse waveform and uncertainty as to the pulse height. In both cases a fraction of the information carried by the pulse has been lost.

3.9 Noise and Spurious Signals

Noise is an approximately random fluctuation superimposed on top of the information content of a signal. All practical electronic signals contain noise, and all practical electronic circuits generate further noise, increasing the overall noise content of a signal as it is processed. The corruption of the information content of a signal as it is degraded by noise becomes more important as the signal level is reduced and it presents an ultimate limit to the lowest levels that can be used.

Resistors (and conductors) generate some noise because of the thermal motion of the electrons. This thermal or Johnson noise is nearly random and contains components uniformly distributed over a wide range of frequencies. A flat spectrum is characteristic of white light, and so this type of noise is termed white noise.

The value of the thermal noise produced by an ideal resistor depends on the resistance, the bandwidth over which the noise is measured and the absolute temperature. Thermodynamic arguments lead to an expression for the noise power generated. This may then be expressed as a RMS voltage

$$v = (4kTBR)^{1/2} \tag{3.3}$$

where R is the resistance, T is the absolute temperature, B is the bandwidth in Hz, and k is Boltzmann's constant. Equation (3.3) yields a noise of about 1 μV measured over a bandwidth of 1 MHz for a 100 Ω resistance at room temperature. This theoretical value is the minimum possible. Practical resistors always show larger values, and in particular produce extra noise when a steady state voltage is applied.

Diodes and transistors generate additional noise as the carriers cross the potential barriers at junctions. This shot noise is wideband and approximately white. Transistors also generate other noise including flicker noise which is confined to the low frequency end of the spectrum and increases approximately as $1/f$. Ideal reactances have no resistive element and generate no thermal noise. However, imperfections in real inductors and capacitors give some noise but normally this is small and is neglected in comparison with the noise sources mentioned above.

Signal to noise ratio

The amount of noise in a signal can be described by the signal to noise (S/N) ratio. This is often expressed on a logarithmic scale in decibel (dB) units. The decibel is widely used in electronics as a ratio measure, and is defined in terms of power. The ratio of two powers p_1 and p_2 is

$$N = 10\,\mathrm{lg}(p_1/p_2)\ \mathrm{dB} \qquad (3.4)$$

where lg indicates logarithm to base ten. If the two powers are dissipated in the same value of resistance, then N can also be expressed in terms of the ratio of the corresponding voltages v_1 and v_2. Then

$$N = 20\,\mathrm{lg}(v_1/v_2)\ \mathrm{dB} \qquad (3.5)$$

because $p = v^2/R$. Although equation (3.5) is only properly valid for voltages measured across identical resistances, it is common to express the ratio of two voltages in decibel units even when this is not the case.

Noise figure

The degradation of the S/N ratio as a signal passes through a circuit or system is measured by the noise figure. This is defined as the ratio between the input and output S/N ratios, where the input signal is assumed to come from an ideal generator and the only input noise is assumed to be thermal noise from the generator resistance. An ideal device, circuit or system contributes no extra noise, and so the noise figure would be unity or zero dB. Practical values range from 10 dB or more down to less than 1 dB for circuits using very low noise devices.

It is still true that the lowest noise figures are achieved from discrete devices rather than integrated circuits. Hence discrete devices are preferred for the best performance with low level signals, although the gap is narrowing as the manufacturing processes of integrated circuits are refined.

Hum and spurious signals

Unwanted components other than the forms of noise described above may be present in a signal or may be added to it as it passes through a system. These spurious signals arise from coupling to other electrical signals nearby. The most common intrusions are hum (50 Hz in the UK) and its harmonics derived from the mains supply. Hum is very difficult to avoid because of the all-pervading presence of mains cables in laboratories and in commercial and industrial premises, and because many electronic circuits derive their power from the mains supply. Careful circuit design, layout and screening can reduce hum levels. Also hum and its harmonics are confined to well defined low frequencies and it may be possible to reduce their effects by filtering to distinguish between the desired and unwanted components in the spectrum of the signal. Filtering is considered further in Chapter 5.

4

Amplification

The amplification of signals is the basis of analogue electronics, and indeed of digital circuitry as well, although this may not be so obvious to the user. Most of the major landmarks in the development of electronics have been associated with the introduction of new or improved amplifying devices. The special property of such devices is that they are active and the input signal modulates power drawn from a supply. Active devices are then able to deliver more signal power than they require as input. This behaviour can be contrasted with that of passive components such as resistors, where the output signals must always be smaller than the input. It is obvious that the property of amplification is essential if we are to detect and process very small signals, but it has many other implications as well.

Very many amplification requirements can be met using integrated circuits. The user need not be concerned with the detailed internal operation of the circuits and very satisfactory design and construction can be done knowing only the properties of the device viewed from the outside. In some cases even these external properties are not critical as we shall see in the case of the operational amplifier. However, for a fuller understanding of how integrated circuits work, and for those applications where they may not be suitable we shall need to understand and exploit the behaviour of the transistor which is used as the basic amplifying element in almost all modern electronic circuits.

4.1 Small Signal Amplification

The term small signal amplification does not mean just the amplification of small signals. It includes a more general concept which leads to considerable simplification in the design of amplifying circuits and is also rather loosely taken to mean amplification where we are concerned more with raising the voltage level of a signal than with developing appreciable signal power. The fundamental concept of small signal amplification is that the operating conditions of the device used remain essentially undisturbed by the presence of the signal. In other words, the signal level is small compared with the quiescent (steady state) currents and voltages. When we are amplifying signals which really are small, such as the microvolts or millivolts that might come from a radio aerial, or from an electrode implanted in a biological specimen, then the assumption is well justified. But even when the signal levels are so large that the assumption is not really valid, the simplifications which result are so valuable that it remains a useful concept.

In Chapter 2 we saw that the behaviour of semiconductor devices—diodes and transistors—is in general non-linear. For example the relation between the base current and the base–emitter voltage of a transistor certainly does not obey Ohm's law, being approximately exponential. However, if we arrange that a transistor has quiescent levels of V_{be} and I_b at the point Q in figure 4.1, and consider only small changes of V_{be} and I_b around Q, the curve in this region can be approximated to a straight line. Then

$$i_b = g_{be} v_{be} \tag{4.1}$$

where g_{be} is the gradient at Q, and lower case symbols for the base current and voltage indicate small changes or signals. We have reduced the relation to a simple straight line. The behaviour can now be described just by a conductance g_{be}, or by a resistance $r_{be} = 1/g_{be}$.

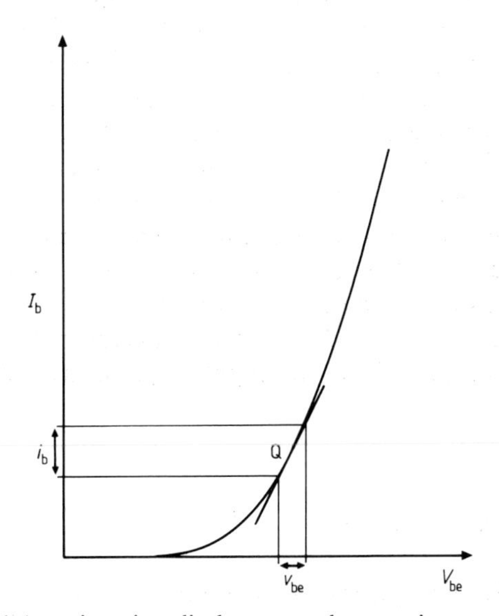

Figure 4.1 A pn junction diode operated at a quiescent point Q.

At first sight this may seem to be a trivial procedure, but the awkward equations coming from the physics of semiconductor devices have been replaced by Ohm's law. Now, circuit calculations which might otherwise occupy a powerful computer for a considerable time can now be done by hand, (or a calculator) in a few moments. In a similar way, other device properties can be reduced to linear relations and represented by simple circuit elements. So a complete amplifier can be treated as a network of linear components, and its behaviour described and analysed with simple mathematics.

4.2 The BPT as an Amplifier

The BPT is a versatile device. It is used extensively in integrated circuits for amplification and switching and can be used by itself—as a discrete component—for the same purposes. MOST and JFET tend to be used rather more exclusively for integrated circuits, although there is an increasing number of discrete applications for which their special properties of low gate current and good isolation between the gate and the source to drain channel make them valuable. We shall concentrate on the use of BPT as small signal amplifiers. The same general principles can be applied to the design of amplifier circuits made with MOST and JFET too. However their very low gate currents allow much higher values of resistance to be used in the input circuits, removing some of the restrictions we shall find with BPT.

Signal amplification is possible provided that the BPT is operated within the active region where transistor action as discussed in §2.3 occurs. Initially it will be adequate to define this region and the properties of the transistor in the region very simply indeed, using an npn transistor as an example. All current and voltage directions are reversed for the pnp version, otherwise the behaviour is the same.

(i) The base–emitter junction is forward biased so that a current I_b flows into the base.

(ii) The collector–emitter voltage is sufficiently positive to ensure that the collector–base junction is reverse biased. It must not be so large that breakdown occurs.

If these two conditions hold then:

(iii) The base–emitter voltage V_{be} is almost constant at about 0.6 V for a silicon transistor.

(iv) The collector current I_c is independent of the collector–emitter voltage V_{ce} and is linearly related to I_b through a constant, the current gain. We shall use the symbol β for the current gain, but α' and h_{FE} are also in common use.

These simplifications are equivalent to approximating the real characteristics of a typical transistor intended for small signal amplification shown in figure 4.2(*a*) by the idealised properties in figure 4.2(*b*).

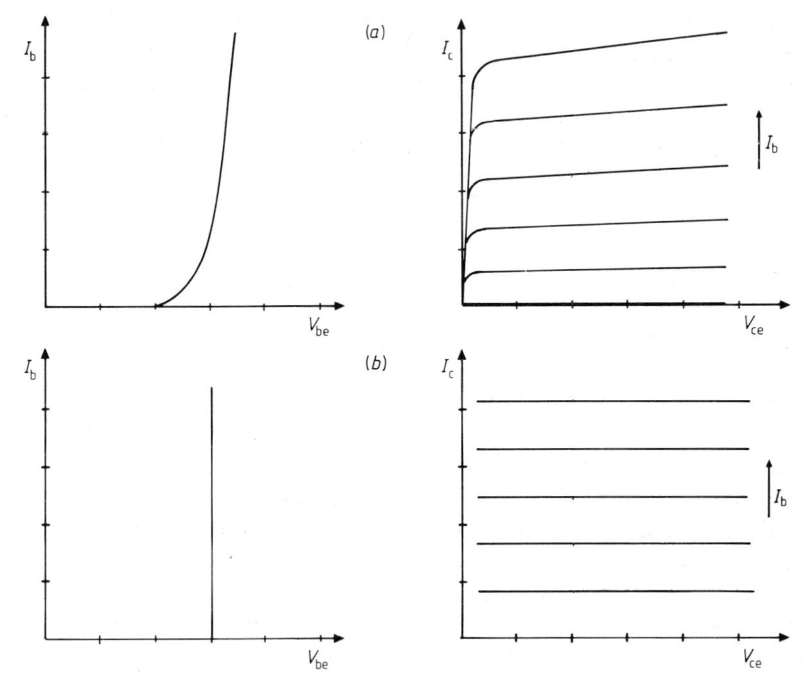

Figure 4.2 The real characteristics of the transistor in (*a*) can be approximated to the idealised characteristics shown in (*b*).

4.3 Biasing

Biasing is the process of establishing the steady state currents and voltages so that conditions (i) and (ii) are satisfied. In most cases it will be necessary to go further and to define the operating conditions quite closely in order to get good performance from an amplifier. It is convenient to consider the biasing process in two parts, separating the input and output requirements.

The most common amplifying circuit injects the input signal into the base of the transistor and extracts an output signal from the collector. The emitter is made the common connection between input and output, (often called ground or earth), and this arrangement is called the common emitter configuration. Other amplifying configurations are possible and we shall return to them later. For the common emitter configuration the two parts of the biasing process correspond to the conditions (i) and (ii) above.

Our initial approach to the input biasing might be to define the base–emitter voltage V_{be} with a voltage source of the value needed to achieve forward conduction. There are three reasons why this is unsatisfactory. First, the exact value of V_{be} required will vary from transistor to transistor.

Second, the base and collector currents change very rapidly with base–emitter voltage. So even if we could set V_{be} accurately to just the right value needed by the particular transistor, I_b would be poorly defined. Finally, the presence of a constant voltage source would short-circuit any signal voltage we attempted to impress on the quiescent value of V_{be}.

A better method is to set the base current and allow V_{be} to adopt whatever value is needed. Probably the simplest current defining circuit is shown in figure 4.3(*a*) where the base current is given by

$$I_b = (V - V_{be})/R_b. \tag{4.2}$$

V_{be} is approximately constant at around 0.6 V, and if V is chosen to be large compared with V_{be}, I_b is very well defined. The input biasing has now determined I_b and V_{be} and also the collector current I_c because

$$I_c = \beta I_b. \tag{4.3}$$

It only remains to set the collector voltage to satisfy condition (ii). Again, if we attempt to define V_{ce} with a perfect voltage source, we shall short-circuit any signal voltage developed at the collector. So the collector is returned to a voltage source through a load resistance R_c as shown in figure 4.3(*b*) then

$$V_{ce} = V_s - I_c R_c \tag{4.4}$$

from Ohm's law. It is normally convenient to use the same voltage source to supply both the base and collector currents and then $V = V_s$.

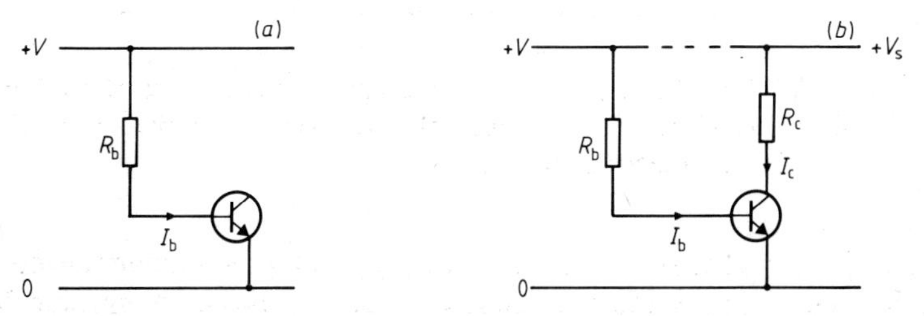

Figure 4.3(*a*) Input biasing of an npn transistor; (*b*) Output biasing of an npn transistor.

These steady state voltages and currents apply in the absence of any signal. If a signal voltage is impressed on the quiescent value of V_{be} then the base current I_b will change. There must then be a corresponding change in the collector current which will develop an output signal voltage across the collector resistance R_c.

A useful way of visualising the operation of this amplifier is to represent its behaviour in graphical form. The series combination of the transistor and load resistance R_c constrains the current through them to be the same, and the sum of their voltage drops must be the power supply voltage V_s. Equation (4.4) is a description of the ohmic behaviour of the load resistance—a plot of I_c against V_{ce} is a straight line intersecting the voltage axis at $V_{ce} = V_s$ and the current axis at $I_c = V_s/R_c$. This load line can be superimposed on a plot of the transistor characteristics as shown in figure 4.4. The quiescent operating point Q of the transistor must lie at the intersection of the load line with the transistor I_c against V_{ce} curve for the chosen I_b.

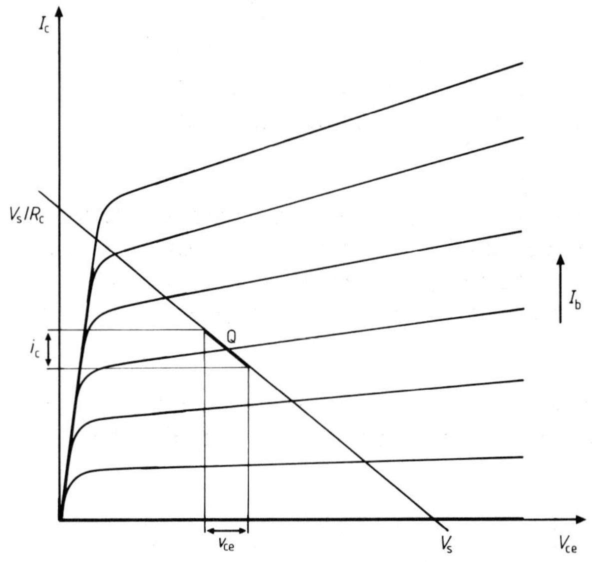

Figure 4.4 A load line plotted on the output characteristics of an npn transistor Small excursions about the quiescent operating point Q produce signal voltage v_{ce} and current i_c.

As the input signal alters I_b, the operating point moves along the load line. The corresponding signals i_c and v_{ce} can be obtained graphically by projection from the current and voltage axes. The relation between the input signal voltage v_{be} and the input signal current i_b can also be obtained graphically by projection from the tangent to the input characteristic as was shown in figure 4.1.

If the output signal level really is small then the precise position of the operating point Q is not important. Almost anywhere along the load line and within the active region of the transistor will probably be satisfactory. But if the collector voltage variation becomes large the transistor will be driven to the extremes of the load line, to the point where the collector current falls to zero (cut-off), and to the knee of the output characteristics (saturation). At these extremes transistor action fails, the collector voltage and current excursions become limited, and the output signal will no longer be linearly related to the input.

Biasing the amplifier so that Q is approximately equidistant from cut-off and from saturation will give the maximum possible swing of the output voltage. Saturation levels of most small signal transistors are well below a volt, so a quiescent value a little above $V_s/2$ is usually chosen. Even when very large output swings are not needed, it is sensible to position Q near the midpoint of the load line to allow for biasing tolerances.

The biasing circuit of figure 4.3 is simple, but has a major disadvantage. Although the base current is very well defined, the collector current and voltage depend directly on the current gain β. This parameter is highly sensitive to variations in the manufacturing process and a range of an order of magnitude is common even for transistors of the same nominal type. An individual transistor will also show some variation of current gain with collector voltage and current and with temperature. A better biasing strategy would be to define I_c and V_{ce} directly rather than indirectly through β.

Figure 4.5 shows an improved biasing circuit which has become almost the standard method for single stage amplifiers. At first sight the circuit may seem to be very similar to the simple arrangement, but its operation is very different. The simple circuit defines an approximately constant base current.

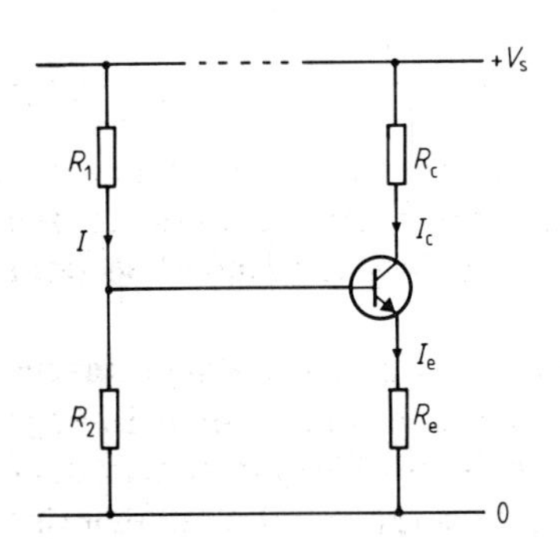

Figure 4.5 Potential divider bias circuit with R_e providing feedback

The improved circuit defines a nearly constant base voltage V_b by potential division from a reference source, usually V_s. So the emitter voltage V_e is fairly well defined because V_{be} is nearly constant. Ohm's law then tells us that the emitter current I_e must be given by

$$I_e = V_e/R_e = (V_b - V_{be})/R_e. \tag{4.5}$$

But $I_e = I_b + I_c$ and $I_c = \beta I_b$. So $I_c = \beta I_e/(1+\beta)$ and is very nearly equal to I_e for large β. V_b is the loaded output from the potential divider R_1 and R_2 and is given by

$$V_b = V_t - I_b R_t \tag{4.6}$$

where V_t and R_t are the Thévenin equivalent output voltage and resistance of the potential divider defined by

$$V_t = R_2 V_s/(R_1 + R_2) \tag{4.7}$$

$$R_t = R_1 R_2/(R_1 + R_2). \tag{4.8}$$

Then the collector current is just

$$I_c = (V_t - V_{be})/(R_e + R_t/\beta). \tag{4.9}$$

For large β and a proper choice of R_1, R_2 and R_e, R_t/β is small compared with R_e. The collector current then becomes almost independent of the current gain. A minor modification of equation (4.4) is needed to allow for the voltage drop across the emitter resistor, so

$$V_{ce} = V_s - I_c R_c - V_e. \tag{4.10}$$

This circuit also compensates for variations in V_{be} from device to device or caused by temperature change. To derive quantitative measures of the stability in each case we can differentiate (4.9) to give

$$\frac{dI_c}{d\beta} = \frac{R_t(V_t - V_{be})}{(\beta R_e + R_t)^2} \tag{4.11}$$

and

$$dI_c/dV_{be} = -(R_e + R_t/\beta)^{-1}. \tag{4.12}$$

Although equation (4.11) cannot be strictly valid for the large variations in current gain often encountered, the results are a useful guide to the effectiveness of the bias arrangement. Good stability is obtained in both cases with large β and R_e and low R_t.

A further stability problem can arise from the collector leakage current which is highly temperature sensitive. Silicon devices have very small leakage currents at room temperatures and this problem is only likely to be serious at very high temperatures. We shall not consider it except to point out that this bias circuit will stabilise the collector current against this effect also.

This circuit is the first example we have met with negative feedback, which is a powerful technique used extensively throughout electronics. What we are really doing is measuring the output of a circuit and comparing it with a reference. Any discrepancy is fed back to the input to compensate for the error and correct the output. In this arrangement the emitter voltage, a measure of the emitter current and hence of the collector current, is compared with $(V_t - V_{be})$. If, for example, the transistor is replaced by one with lower current gain then I_c, I_e and V_e tend to fall. But if V_t is held constant, a decrease in V_e must increase V_{be} and hence I_b. So I_c recovers to a level only a little below the original value.

Calculation of the operating point and stability of a given circuit is easy using equations (4.7)–(4.11). Designing a circuit to bias a transistor at a particular collector current and voltage is not quite so straightforward because there are compromises to be made. A high ratio of R_e/R_t is desirable for good stability. However, if R_e is made very large the voltage drop across it reduces the maximum output voltage swing available according to equation (4.10). R_1 and R_2 are effectively shunted across the amplifier input and low values may degrade the input signal.

As an example, let us suppose that we wish to calculate suitable values for R_1, R_2, R_e and R_c in figure 4.5 to set the operating point of the transistor at $I_c = 2$ mA and $V_{ce} = 7$ V. The power supply is assumed to be 12 V and the transistor to have a minimum current gain of 100. A useful starting point is to allow about 1 V across R_e. This will make V_e large enough to swamp any likely variations in V_{be} so that $(V_t - V_{be})$ is reasonably constant. At the same time it is not so large that the output voltage swing is seriously reduced. So $V_e = 1$ V, $R_e = 500\ \Omega$ and $V_b = 1.6$ V if we take V_{be} to be 0.6 V.

For the worst case (minimum) of a current gain of 100, R_t can be about 10 times R_e before the R_t/β term in equation (4.9) becomes significant. Since V_t is a small fraction of V_s, R_1 will be very much larger than R_2 and so R_t is only a little less than R_2. So we may choose $R_2 = 10R_e = 5\ \text{k}\Omega$. Then R_1 must be $R_2(12 - 1.6)/1.6 = 32.5\ \text{k}\Omega$. Finally R_c is $(12 - 7 - 1)/2\ \text{mA} = 2\ \text{k}\Omega$. In practical circuits using discrete devices and components we may have to use the nearest preferred resistor values rather than the figures derived above.

Substitution of these resistance values into equations (4.11) and (4.12) gives the stabilities of I_c to changes in β and V_{be} as 1.5×10^{-6} and 1.8×10^{-3}, respectively. An increase in the current gain from 100 to 150 or a decrease in V_{be} of 50 mV would increase I_c by less than 5%. Therefore this is a fairly stable design.

An alternative way of choosing values for R_1 and R_2 is to allow sufficient standing current down the potential divider that we may take the base voltage V_b to be the same as the unloaded output V_t. A standing current of ten times the maximum expected base current will probably be adequate. Then, $I = V_s/(R_1 + R_2)$ combined with equation (4.7) will give R_1 and R_2.

The presence of negative feedback through R_e stabilises the collector current and voltage against disturbances whatever the cause. So in addition to providing stability against device parameter changes, our circuit will also attempt to nullify the effect of any input signals. The gain of the amplifier will therefore be reduced. However, in most cases the changes in device parameters are long term, whereas signals are short term variations. We can restore the lost gain if the circuit is modified to remove feedback for rapid variations. But the feedback and stability must be retained for steady state and slowly changing quantities.

Figure 4.6 shows a capacitor bypassing the emitter of the transistor to the common line. The value of C_e is chosen so that its reactance is very small at signal frequencies, virtually a short-circuit, so there can be no signal voltage developed across it, no feedback and no gain reduction. For steady state and slowly varying quantities the reactance of C_e becomes very large, and R_e provides feedback and stability.

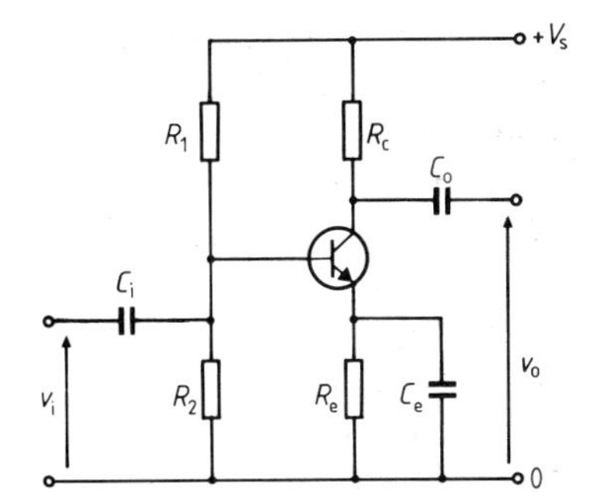

Figure 4.6 A single stage voltage amplifier using an npn transistor.

Our amplifier is not quite complete. The input signal may come from a source with a resistive path to the common line. If so, that path will act as a shunt to the output of the potential divider, alter V_b, and disturb the bias conditions. Again, a capacitor can be used to distinguish between steady state and signal quantities. The capacitor C_i blocks any resistive path but has a low reactance at signal frequencies and so signals can be coupled to the base of the transistor. A similar coupling capacitor may be needed at the amplifier output if an external load has a resistive path to ground.

Bypass and coupling capacitors cannot be used in amplifiers required to operate down to extremely low or zero frequency. More complicated circuits, often using many active devices, are then needed to achieve the required gain and to allow direct coupling throughout the amplifier.

4.4 Transistor Models

In the bias calculations above we approximated the real properties of a transistor by idealised elements such as voltage and current sources. Then we were able to apply simple arguments using Ohm's law to deduce the behaviour of the circuit. Rather more formally, we could have produced a network model of the device.

Our highly simplified description of a transistor in §4.2 corresponds to the model for the steady state properties shown in figure 4.7(*a*). The base–emitter voltage is represented by the voltage source, and the relation between I_b and I_c is modelled by a current-controlled current source. This description is so simple that the model is not essential for the bias calculations, but more sophisticated models are invaluable for analysing the performance of circuits.

A more accurate steady state model, shown in figure 4.7(*b*), includes a resistance R_{be} in series with the input voltage source to account for the increase of V_{be} with I_b. A shunt resistance R_{ce} across the current source accounts for the increase in I_c as V_{ce} is raised. Then, $V_{be} = 0.6\ \text{V} + I_b R_{be}$ and $I_c = \beta I_b + V_{ce}/R_{ce}$.

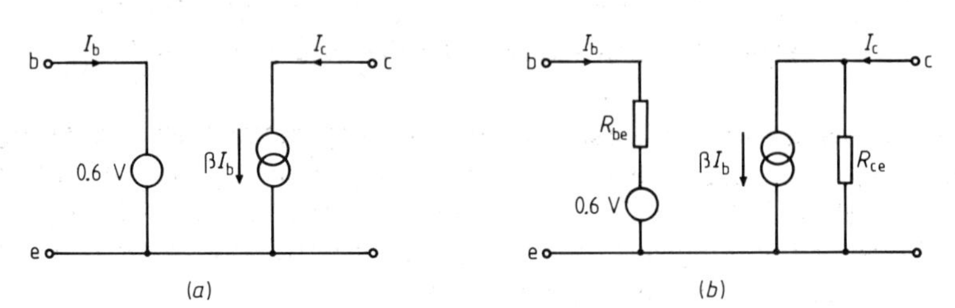

Figure 4.7(*a*) A very simple model for the steady state behaviour of a transistor; (*b*) a more accurate model.

A similar model for the small signal behaviour as shown in figure 4.8 will enable us to calculate the signal voltage gain of an amplifier. A constant voltage source cannot have any signal voltage across it, so the transistor input is modelled just by a resistance r_{be} representing the dynamic resistance of the base–emitter junction at the operating point as indicated in equation (4.1). Transistor action can be represented by a current generator controlled by the signal base current. The signal current gain i_c/i_b is the slope of the I_c versus I_b relation measured at constant V_{ce} and is approximately equal to the steady state current gain. We shall use the same symbol β for both. A shunt resistance $r_{ce} = v_{ce}/i_c$ models the dynamic slope of the output characteristic, I_c versus V_{ce}.

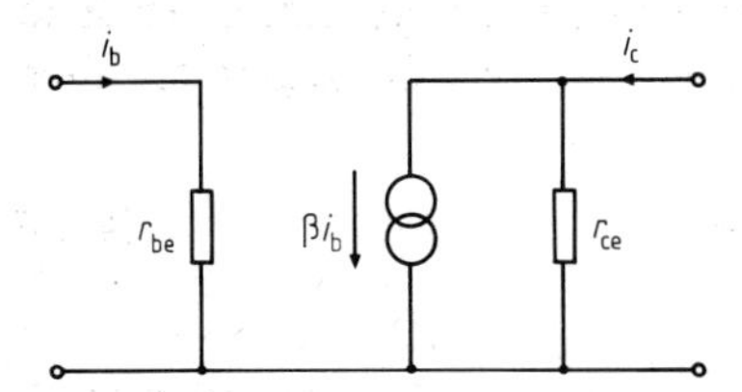

Figure 4.8 A current controlled model of the signal behaviour of a transistor.

4.5 Voltage Gain

Derivation of the voltage gain is straightforward. The device model is combined with the other circuit components to form the complete network shown in figure 4.9. Note that the power supply, assumed to be a constant voltage, cannot have any signal voltage across it and so is modelled with a short-circuit. The emitter bypass capacitor and the input and output

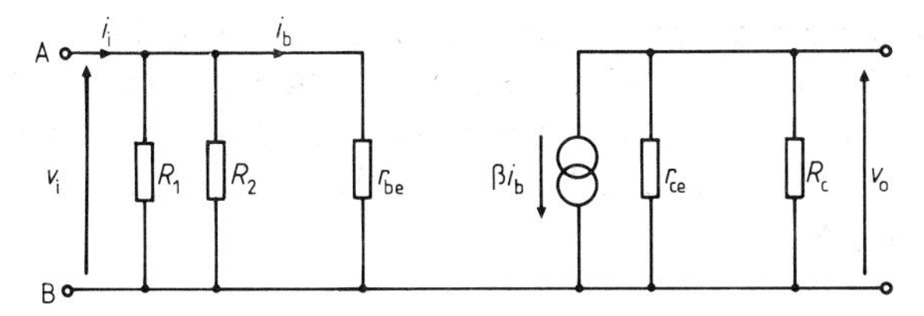

Figure 4.9 Complete model for the single stage amplifier.

coupling capacitors are assumed to have negligible reactance at signal frequencies and are also replaced by short circuits. It is convenient to combine R_1 and R_2, replacing them by their parallel resultant R_t. R_c can be combined with r_{ce} to give R_c' which further simplifies the network. Then we can write for the input and output loops.

$$v_i = i_b r_{be} \tag{4.13}$$

$$v_o = -\beta i_b R_c'. \tag{4.14}$$

Eliminating i_b yields the voltage gain

$$A_v = v_o/v_i = -\beta R_c'/r_{be}. \tag{4.15}$$

The minus sign indicates that the sense of the output signal is inverted with respect to the input which corresponds to a phase shift of 180° or π for a sinusoidal signal.

An equivalent model for the behaviour of the transistor can be based on the view of a BPT as a voltage controlled device. Figure 4.10 shows a model with the generator producing a current $g_m v_{be}$ where g_m is a transconductance i_b/v_{be}. Again, r_{be} and r_{ce} model the slopes of the input and output characteristics. The current controlled and voltage controlled models are equivalent and so equating the outputs of their current generators leads to $\beta = g_m v_{be}/i_b = g_m r_{be}$. An alternative expression for the voltage gain can then be obtained from equation (4.15)

$$A_v = -g_m R_c'. \tag{4.16}$$

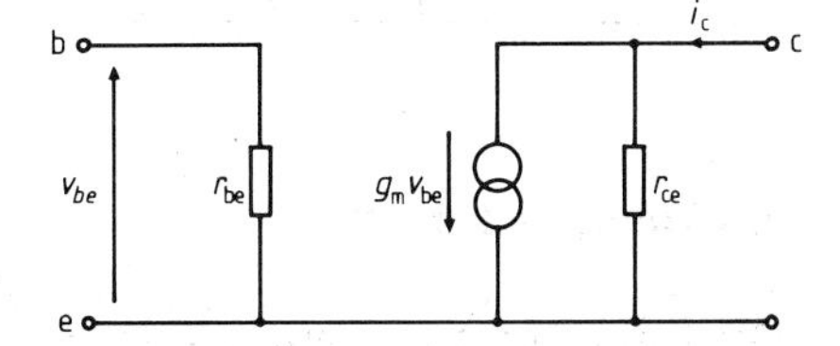

Figure 4.10 Voltage control model for a transistor.

In subsequent analysis of amplifier properties we shall concentrate on the current controlled model. However, equivalent expressions can be derived using the transconductance model, or obtained by substituting $\beta = g_m r_{be}$.

4.6 Input and Output Resistance

Neither R_1 nor R_2 have any influence on the voltage gain of the amplifier when it is driven from a signal source with negligible output resistance. They do however affect the input resistance looking into the terminals A and B in figure 4.9 because they appear directly shunted across the input. So the input resistance r_i, which is the ratio of the input voltage to the input current, will be the parallel combination of R_1, R_2 and r_{be}.

A safe definition of the output resistance of a network is the ratio of the open-circuit output voltage to the short-circuit output current. We already have the open-circuit output voltage in equation (4.14) and a short-circuit across the output will sink the entire current from the controlled source. So the output resistance is

$$r_o = -\beta i_b R_c' / -\beta i_b = R_c'. \tag{4.17}$$

In this case, r_o is just the resistance looking back into the amplifier with the current source suppressed. Most amplifiers will deliver their output into an external load R_l as shown in figure 4.11. The loaded voltage gain A_v' can be deduced by including the effect of the potential divider formed by r_o and R_l, so that

$$A_v' = A_v R_l / (R_l + r_o). \tag{4.18}$$

Alternatively the external load can be combined with R_c' so that the total shunt load across the current generator R_l' is the parallel combination of r_{ce}, R_c and R_l. Substituting R_l' for R_c' in equation (4.15) then leads to the same expression for the loaded voltage gain.

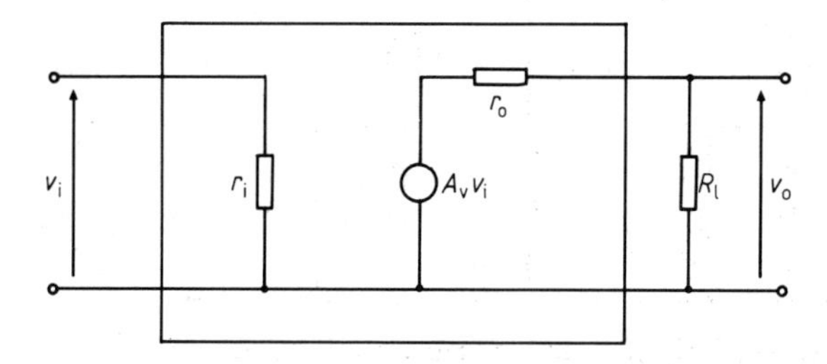

Figure 4.11 Model for an amplifier with an external load R_l.

Current and power gain

Although we are most often concerned with the voltage gain, the current and power gains of a small signal amplifier are also of interest. The input current $i_i = v_i / r_i$ and the output current into the external load is $i_o = A_v' v_i / R_l$ and so the loaded current gain A_i' will be

$$A_i' = i_o / i_i = -\beta r_i R_l' / r_{be} R_l \tag{4.19}$$

from equation (4.15). The power gain is the product of the voltage and current gains.

A transistor with a current gain of 100 operating at a collector current of 2 mA will typically have $r_{be} = 1.25$ kΩ and $r_{ce} = 10$ kΩ. Substituting these values into equation (4.15) gives an open circuit voltage gain of − 133 for the amplifier in figure 4.6. The input and output resistances are 970 Ω and 1.67 kΩ from equations (4.17) and (4.18) respectively. These results apply over the range of signal frequencies where capacitor reactance is negligible and where the device and component modelling is valid.

4.7 Multistage Amplifiers

The value of − 133 we obtained for the voltage gain of this single stage amplifier is fairly typical. If we require more gain it will be necessary to add further stages of amplification. For example, we could cascade two of our single stage amplifiers to form the two stage amplifier shown in figure 4.12. Only one coupling capacitor is needed to obtain isolation between the two stages. Although the stages are identical, the input of the second stage loads the output of the first and reduces its gain. The loaded gain of the first stage is − 49 from equation (4.18), giving an overall gain of just over 6500.

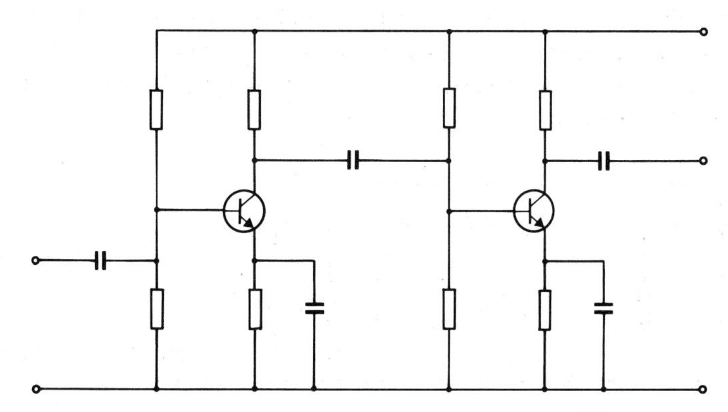

Figure 4.12 Two capacitively coupled voltage amplifier stages.

Each transistor in figure 4.12 is separately biased and the steady state conditions in the two stages are isolated by the coupling capacitor. With care it is possible to dispense with the coupling capacitor and allow direct coupling between the stages. Then the biasing of both transistors must be achieved simultaneously.

Figure 4.13 shows a directly coupled two stage amplifier which has overall negative feedback to establish the operating conditions of both transistors. As before, we shall simplify the bias analysis by assuming that the two

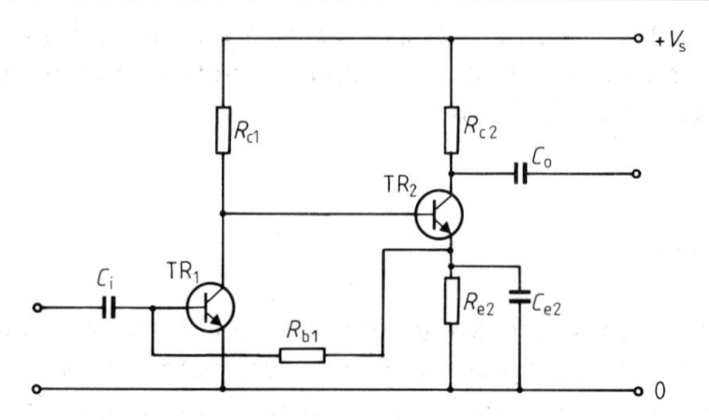

Figure 4.13 A two-stage directly coupled amplifier.

transistors have high current gains and that V_{be} is approximately constant at about 0.6 V. Then

$$V_{e2} = V_{b1} + I_{b1}R_{b1}$$
$$= V_{b1} = 0.6 \text{ V} \tag{4.20}$$

provided that R_{b1} is not too large. The emitter current of the second transistor is then

$$I_{e2} = V_{e2}/R_{e2} = 0.6 \text{ V}/R_{e2} = I_{c2}. \tag{4.21}$$

So the collector current of the second stage has been defined using the base–emitter voltage of TR_1 as a reference. Now the collector voltage of TR_2 is defined through the voltage drop in the collector resistor.

$$V_{c2} = V_s - I_{c2}R_{c2}. \tag{4.22}$$

Direct coupling between the stages forces the collector voltage of TR_1 to be the same as the base voltage of TR_2. Assuming again that $V_{be2} = 0.6$ V, the operating conditions of TR_1 must be

$$V_{c1} = V_{b2} = 1.2 \text{ V} \tag{4.23}$$

and

$$I_{c1} = (V_s - V_{c1})/\text{R}_{c1}. \tag{4.24}$$

A collector voltage of 1.2 V is low and TR_1 may be close to saturation. Only small voltage excursions are possible, but the high gain of the second stage means that the signal level here must be small, so this is not a serious limitation.

Two comparisons are being made in this negative feedback circuit. V_{e2} is being compared with V_{b1} to set I_{e2} and hence I_{c2}. Less obviously, V_{c1} is being compared with $V_{be1} + V_{be2}$ to control I_{c1}. In either case the feedback will act to maintain the two currents at these values. Very high stability against current gain variations is achieved, but the circuit is sensitive to the base–emitter voltages which are used as references.

Signal feedback will also be present through R_{b1} unless V_{e2} is decoupled with a shunt capacitor. Input and output coupling capacitors will also normally be needed for isolation.

4.8 Buffer Stages

An ideal voltage amplifier would have infinite input resistance giving negligible loading on the signal source, and zero output resistance allowing constant voltage drive into any external load. Practical amplifiers are not ideal and loading effects will be present as in the signal transfer between the stages of the circuit in figure 4.12.

For signal sources which have a very high output resistance it is desirable to have amplifiers with large values of input resistance to avoid severe loading. This can be achieved at the expense of voltage gain by the circuit in figure 4.14. (Bias and coupling arrangements have been omitted for clarity.) By analogy with the common emitter amplifier this should be called a common collector circuit but it is more usually known as an emitter follower. Since V_{be} is nearly constant, the emitter voltage 'follows' an input voltage applied to the base and the voltage gain is close to unity. An analysis of the circuit using the simple transistor model of figure 4.8 will confirm this result.

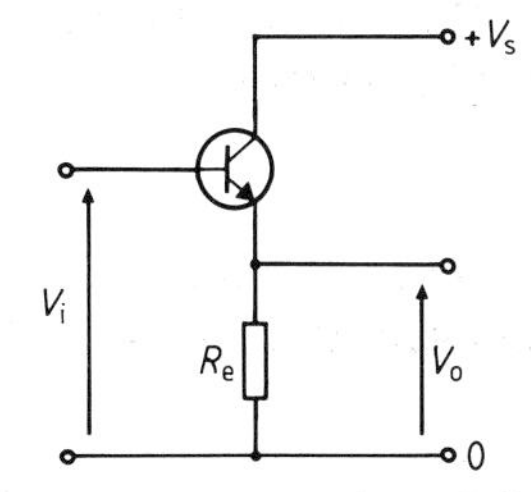

Figure 4.14 The emitter follower.

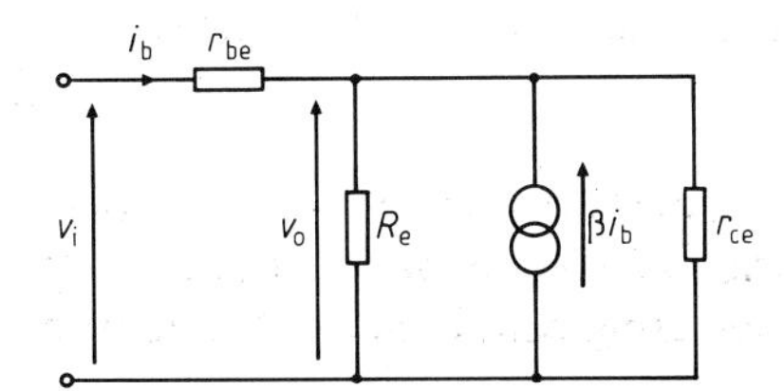

Figure 4.15 Model for the emitter follower.

Figure 4.15 shows the network obtained by combining the transistor model with the remainder of the circuit. Writing expressions for the input and output loops gives

$$v_i = i_b r_{be} + v_o \tag{4.25}$$

$$v_o = (i_b + \beta i_b) R'_e \tag{4.26}$$

where R'_e is the parallel combination of R_e and r_{ce}. Solving for the voltage gain yields

$$A_v = \frac{v_o}{v_i} = \frac{(1+\beta)R'_e}{r_{be} + (1+\beta)R'_e} \tag{4.27}$$

which is very close to unity for high current gain transistors and reasonable values of R_e, r_{be} and r_{ce}. The input resistance is the ratio of the input voltage to the input current and so from equations (4.25) and (4.26)

$$r_i = r_{be} + (1+\beta)R_e' \tag{4.28}$$

which can be very large if the current gain β is high.

Examination of figure 4.15 shows that the base current becomes v_i/r_{be} when the output is short circuited. So the short circuit output current is $(1+\beta)v_i/r_{be}$ and the output resistance is just

$$r_o = r_{be}/(1+\beta) \tag{4.29}$$

if we take the open circuit voltage gain to be unity. r_o can be very low if β is large. This result assumes negligible source resistance. For the more realistic case where the source resistance r_s is appreciable, r_{be} in equation (4.29) can be amended to $r_{be}+r_s$.

The emitter follower can therefore be used as a buffer between a high resistance source and a low resistance load. The unity voltage gain, high input resistance and low output resistance reduce signal loss from loading effects. However, bias arrangements will normally be needed and adding the potential divider of the standard bias circuit as shown in figure 4.16, reduces the overall input resistance. R_1 and R_2 cannot be made too large before the loading of the potential divider by the base current upsets the bias conditions and degrades the stability. In order to achieve high overall input resistance the device must have an extremely low input current.

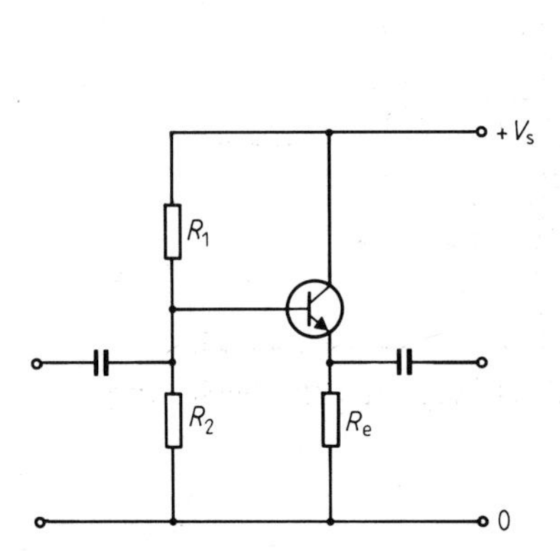

Figure 4.16 Emitter follower with potential divider bias.

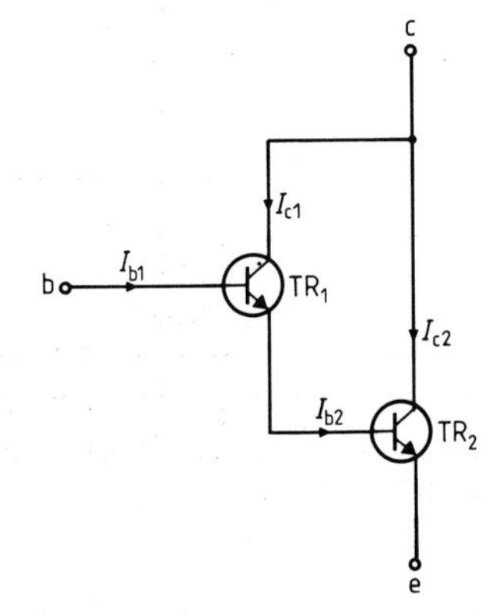

Figure 4.17 A Darlington pair.

Commonly available transistors rarely have current gains above a thousand. But it is possible to produce a compound 'transistor' with very high effective current gain by cascading transistors. Figure 4.17 shows two npn transistors, TR_1 and TR_2 with gains β_1 and β_2 connected so that the emitter current of TR_1 forms the base current of TR_2. Then

$$I_{c2} = \beta_2 I_{b2} = \beta_2(1+\beta_1)I_{b1}. \qquad (4.30)$$

So the effective current gain of the compound transistor $(I_{c1}+I_{c2})/I_{b1}$, is very nearly the product of the current gains of the individual transistors. This arrangement, known as a Darlington pair, can be extended to more than two transistors if desired.

The MOST has a very low input current and is useful here. The basic input resistance of the device itself can be very large, perhaps as high as $10^{12}\ \Omega$. Bias arrangements will reduce this, but very high resistor values can be used in the bias chain, and an overall input resistance of 100 MΩ or more is readily achieved. JFET also have high input resistances, and can be used for slightly less critical applications.

In some cases it may be possible to use direct coupling and reduce or eliminate the need for extra bias components. For example, an emitter follower can easily be added to the feedback pair amplifier of figure 4.13 to provide a very low output resistance. The voltage on the collector of the second transistor will presumably have been chosen at a little above $V_s/2$ to allow a reasonable output voltage swing. This will be ideal for the base voltage of the emitter follower, defining the emitter voltage at about $V_s/2$. Therefore direct connection as shown in figure 4.18 is possible. The emitter current is defined through $I_e = V_e/R_e$ as for the standard bias circuit.

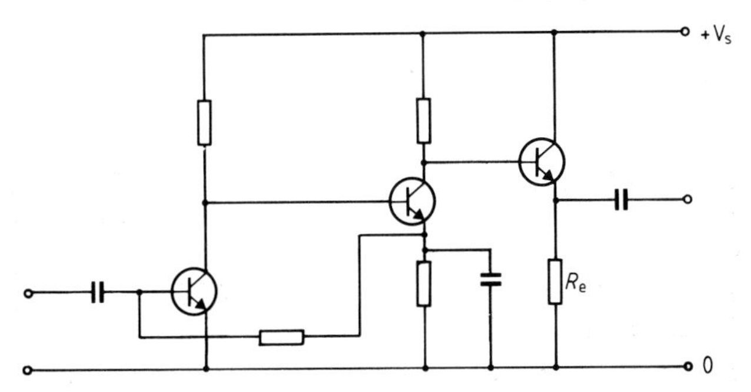

Figure 4.18 Emitter follower output added to the directly coupled two-stage amplifier.

4.9 Frequency Response

At sufficiently low frequencies the reactance of bypass and coupling capacitors will become appreciable and can no longer be neglected. The effect of each capacitor on the gain can be considered separately. C_i and r_i in the single stage amplifier of figure 4.6 act as a potential divider as shown in figure 4.19(*a*), reducing the signal reaching the base of the transistor at low frequencies and hence reducing the gain (they form a high pass filter as discussed later in §6.1). It is convenient to take the case where the capacitor reactance is numerically equal to the resistance as defining the frequency at which the effect becomes important. Well above this transition or corner frequency the signal loss is negligible, below it the attenuation and gain reduction increase steadily as shown in figure 4.19(*b*). The reactance of a capacitor is $1/\omega C$ where ω is the angular frequency $2\pi f$, and so the corner frequency is given by

$$f_l = (2\pi C_i r_i)^{-1}. \tag{4.31}$$

The output capacitance C_o is in series with output resistance r_o and R_l and leads to a similar low frequency limit at $[2\pi C_o(r_o + R_l)]^{-1}$. There is a 3 dB signal loss at a corner frequency, and depending on the gain accuracy required, it may be necessary to place these corner frequencies well below the lowest signal frequency. A value of 10 μF for C_i gives a corner frequency of 16 Hz and would be suitable for signals down to perhaps 50 Hz. A similar value might be suitable for C_o.

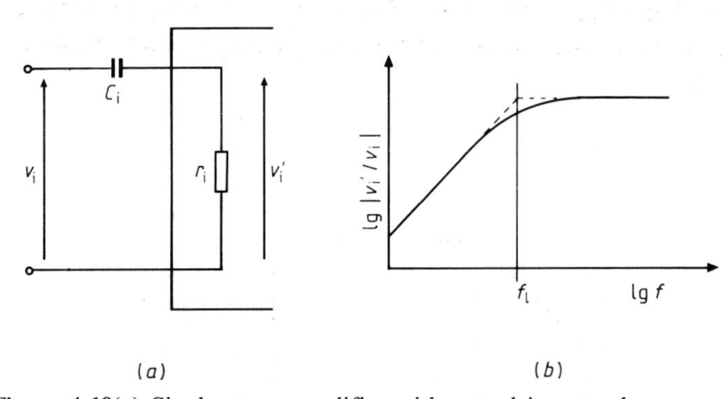

Figure 4.19(*a*) Single stage amplifier with actual input voltage v_i and effective input voltage v_i', having been modified by the input coupling capacitor c_i and input resistance r_i. (*b*) Corresponding attenuation v_i'/v against frequency f on logarithmic scales.

The emitter bypass capacitor C_e also imposes a low frequency limit. C_e bypasses the output resistance at the transistor emitter. This will be very nearly the same as the output resistance of an emitter follower. In our example, the output resistance is a little less than 12.5 Ω so a capacitor of about 1 mF is needed to maintain good bypassing for signal frequencies down to 50 Hz. The overall low frequency response includes all of these three effects, but usually one of them is predominant and it determines the frequency at which the gain begins to fall.

High frequency limitations

Transistor action falls at high frequencies. When the time taken for carriers to cross the base region becomes comparable with the signal period, the current gain is reduced. Device manufacturers often specify the cut-off frequency f_T at which the current gain of a transistor has fallen to unity. However, this figure can be deceptive. Amplifier performance will be affected at frequencies well below f_T.

High frequency performance is limited by other effects. Signal feedback from the collector to the base through the depletion layer capacitance of the reverse-biased junction reduces the voltage gain of an amplifier at high frequencies. Stray shunt capacitance and series inductance in the circuit will also become important at high frequencies and thus the gain will be further reduced. The effects of strays can be reduced and device performances can be improved by reducing the physical dimensions. It is possible to fabricate small transistors which have cut-off frequencies of several gigahertz, and with careful circuit design useful gain can be achieved even at frequencies of above a gigahertz.

The single stage amplifier (figure 4.6) that has been considered might use a general purpose small signal discrete transistor with a cut-off frequency of 100 MHz. Then the upper corner frequency f_u is likely to occur at a few hundred kilohertz and the overall frequency response have the form shown in figure 4.20.

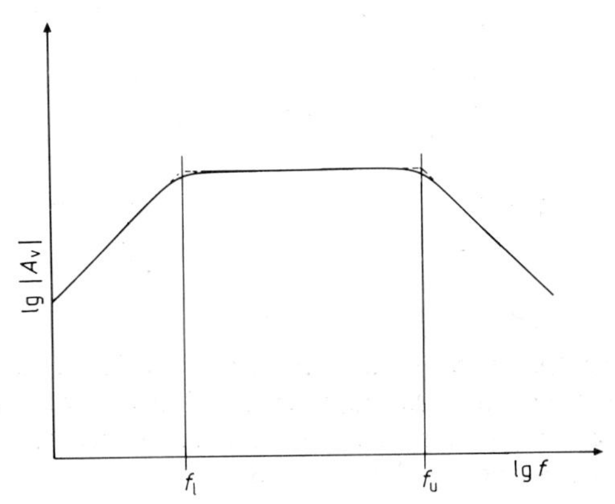

Figure 4.20 Overall frequency response with lower and upper frequency limits, f_l and f_u.

4.10 The Operational Amplifier

Although small signal amplifiers are made from discrete transistors for special purposes, integrated circuit amplifiers are used for the majority of applications. An integrated circuit amplifier may contain many transistor amplifier stages, but often costs no more than a single stage transistor amplifier made from discrete components and thus it is sensible to use them wherever possible. Integrated circuit amplifiers are available to suit almost every conceivable application, but the most general type, the operational amplifier, can be tailored to perform a wide variety of functions.

The internal circuitry of an operational amplifier incorporates one or more stages similar to that shown in figure 4.21(*a*). The symmetry of this circuit implies that the output voltage V_o must be zero for $V_1 = V_2$ if the two transistors are identical and if they drive identical load resistors. This will remain true even if, say, their base–emitter voltages change with temperature, provided that both transistors experience the same temperature change. Likewise, the output voltage is not dependent on the supply voltages or on the absolute values of V_1 and V_2. The symmetry of the circuit is disturbed when $V_1 \neq V_2$, and then an output voltage is generated which is proportional to the difference between the input voltages.

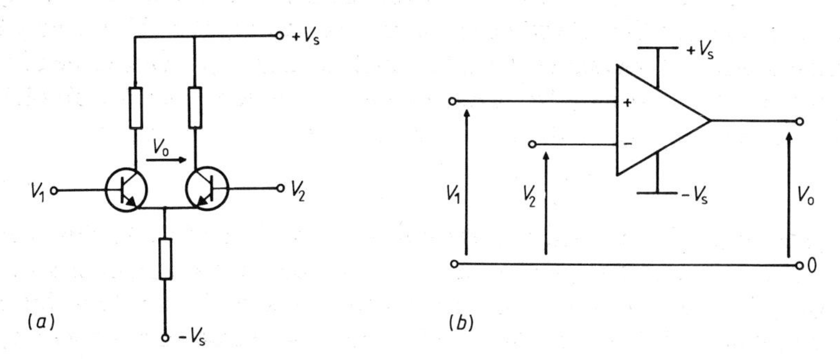

Figure 4.21(*a*) A differential amplifier; (*b*) an operational amplifier.

Proper operation of this differential amplifier requires accurate matching of the two transistors and their load resistors. This is difficult to achieve with discrete components, but is relatively easy in an integrated circuit.

An operational amplifier is a special case of a differential amplifier (figure 4.21(*b*)). The output V_o is proportional to the difference between the two input signals, V_1 and V_2, that is

$$V_o = A_v(V_1 - V_2) \qquad (4.32)$$

where A_v is the differential voltage gain. Upper case symbols are used here because the amplifier can operate with steady state quantities, but the equation is valid for signals too. Note that the amplification from the V_1 input is positive, whereas that from the V_2 is negative, corresponding to inversion of the signal. It is usual to use + and − signs to distinguish the non-inverting and inverting inputs and it is important to remember that these symbols refer to the sign of the gain and not to the sign of the signal. Power supplies, usually symmetrical about the common line, are required as shown in figure 4.21, but it is common to omit these connections from circuit diagrams.

The other particular feature of the operational amplifier is that the voltage gain A_v is very large. Further characteristics include high input resistance and low output resistance, but these are not essential. Negative feedback is then usually applied to define the function of the circuit. In figure 4.22, R_f provides feedback from the output to the inverting input, and this with R_i defines the overall voltage gain. For the input and output loops

$$V_i = I_i R_i + V \tag{4.33}$$

$$V_o = I_f R_f + V \tag{4.34}$$

and at the inverting input

$$I = I_f + I_i \tag{4.35}$$

where V and I are the voltage and current at the inverting input. Now the gain A_v is so large that V will be small compared with V_i or V_o and may be neglected. Also, if V is small, I will be small for any reasonable value of the input resistance and may be neglected too. Then equations (4.33)–(4.35) may be combined to give the voltage gain of the complete circuit

$$V_o/V_i = -R_f/R_i. \tag{4.36}$$

Hence the overall gain is simply determined by the ratio of the feedback and the input resistors. The precise value of the operational amplifier's basic voltage gain is unimportant, provided only that it is sufficiently large for the approximations to be valid. This circuit, where the overall gain is negative, is called the inverting configuration.

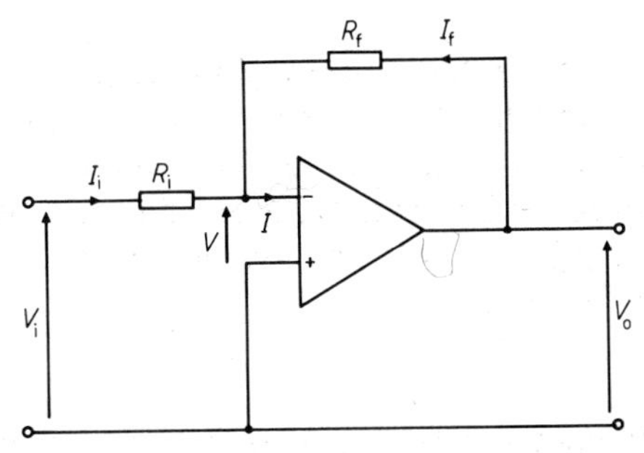

Figure 4.22 The inverting configuration of an operational amplifier.

In figure 4.22 the non-inverting input is connected to the common line—colloquially it is 'earthed'. If V is small then the inverting input is also effectively at earth potential, although not directly connected to earth. This point is called a virtual earth, and the concept is often invoked to simplify the analysis of operation amplifier circuits, bypassing the explicit equating of V and I to zero.

A rearrangement of figure 4.22 where the input signal is applied to the non-inverting input gives the amplifier configuration shown in figure 4.23. Again we can write equations for the input and output voltages

$$V_i = I_f R_i \tag{4.37}$$

$$V_o = I_f(R_i + R_f) \tag{4.38}$$

where we have already neglected the amplifier input voltage and current. Then the overall voltage gain is

$$V_o / V_i = (R_i + R_f)/R_i = 1 + R_f/R_i. \tag{4.39}$$

Application of the virtual earth principle to figure 4.22, or examination of equation (4.33), shows that the input resistance of the inverting configuration must be just R_i. For the non-inverting amplifier the input resistance is very large—almost infinite. For both configurations the output voltage is independent of any current drawn by an external load, corresponding to a perfect voltage source of negligible output resistance.

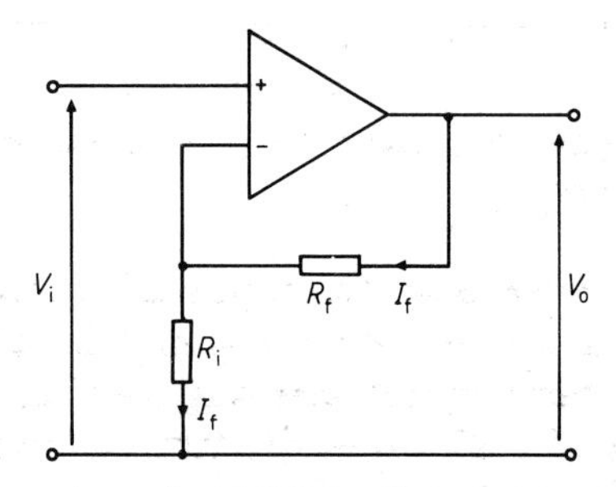

Figure 4.23 The non-inverting configuration of an operational amplifier.

A special case of the non-inverting configuration occurs when either R_f is zero or R_i is infinite (or both). Then the gain is unity and the output voltage follows the input voltage. This voltage follower is an operational amplifier equivalent to the emitter follower, and is an almost perfect buffer.

The results derived above apply to ideal operational amplifiers. Practical devices will have limitations which may need to be taken into account. The differential voltage gain is not infinite. Values of 10^4–10^5 are commonly achieved. However, provided that the overall voltage gain with feedback is limited to a small fraction of the gain without feedback, the errors introduced by the virtual earth approximation are very small.

The high frequency response without feedback is limited because of inherent high frequency effects in amplifiers mentioned earlier. Many operational amplifiers also have a severely restricted response as a deliberate feature to allow operation with a wide range of feedback components. A typical frequency response is shown in figure 4.24, where it will be seen that the low frequency gain is maintained up to only about 10 Hz. However, the application of feedback raises the corner frequency as indicated on the lower plot, and operation up to high frequencies is possible if the final gain is restricted.

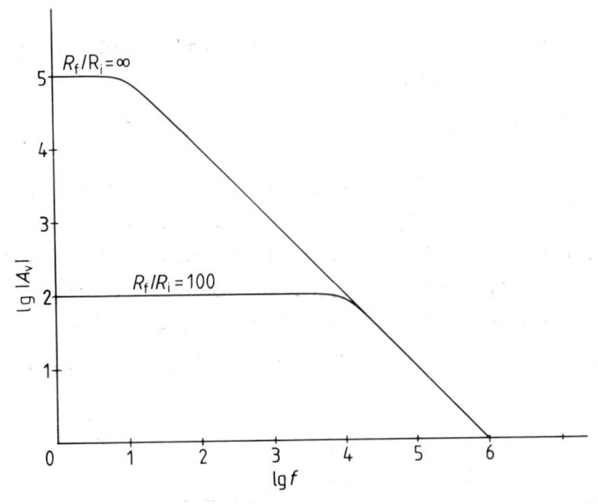

Figure 4.24 Frequency response of a typical operational amplifier with and without feedback.

The output voltage range is limited as with all practical amplifiers. If the amplifier is driven into saturation the gain falls and the virtual earth approximation fails. Most operational amplifiers use symmetrical power supply voltages, often of + 15 V and − 15 V and allow the output voltage to swing to within about a volt of these levels.

The currents drawn by the input terminals are not zero. As well as the signal current there will be a quiescent bias current. Operational amplifiers with bipolar input stages may draw input currents of a few 100 nA, which might need to be taken into account if high values of R_i or R_f are used. Alternatively, operational amplifiers with MOST or JFET input stages are available and these have bias currents in the picoamp range. Then very high values of circuit resistance can be used without introducing any significant error.

5

Power Amplification and Power Supplies

The transistor and operational amplifier circuits in Chapter 4 are intended to operate on voltage signals and can deliver only small amounts of power into a load. An emitter follower stage can drive a signal into relatively low loads, and may generate a few tens of volts across a few hundred ohms. This corresponds to a power level of up to about 1 W. If we require higher levels, as when interfacing to electromechanical devices such as motors and loudspeakers, we need circuits which are specifically designed to generate appreciable quantities of electrical power. The signal excursions will be large, consequently the small signal approximation may fail. We shall also be concerned with the efficiency of converting power drawn from the power supply to signal power developed in the load. As with the small signal amplifiers in Chapter 4, we shall use BPT in the circuit examples. However, MOST are becoming popular for power amplification because they require very little input current and input power. Also, they are less likely to suffer catastrophic damage when overloaded.

5.1 Efficiency and Class A Operation

A value for the efficiency of a basic single transistor amplifier can easily be derived. To simplify the problem we shall disregard details of the bias arrangements, and assume an idealised BPT which can operate down to $V_{ce}=0$, has a negligible V_{be} and a high current gain. No matter which configuration of the transistor amplifier is chosen the efficiency turns out to be the same. Here, an emitter follower driving directly into a load resistor R_l, as shown in figure 5.1, is used as the example. A plot of V_o versus V_i, the transfer characteristic, is shown in figure 5.2 for this idealised circuit. The voltage gain is approximately unity between cut-off where $V_o=0$ and saturation where $V_o=V_s$, and falls abruptly to zero outside that range.

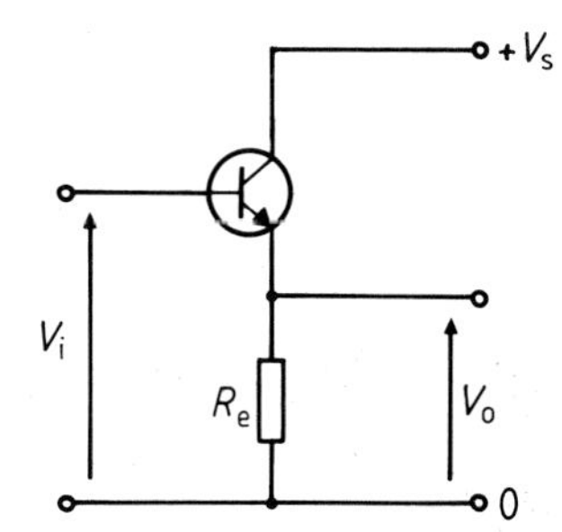

Figure 5.1 An emitter follower as a power amplifier.

It is usual, although not often realistic, to assume that the input waveform is sinusoidal. To obtain the maximum possible undistorted output power we require the maximum possible output voltage swing without driving the amplifier beyond the linear region. This will be achieved if the quiescent emitter voltage, V_e, is set at $V_s/2$ allowing an excursion from cut-off to saturation. The output swing is V_s peak to peak or $V_s/2\sqrt{2}$ RMS, corresponding to an output signal power developed in the load resistance R_e given by

$$P_o = (V_s/2\sqrt{2})^2/R_e = V_s^2/8R_e. \tag{5.1}$$

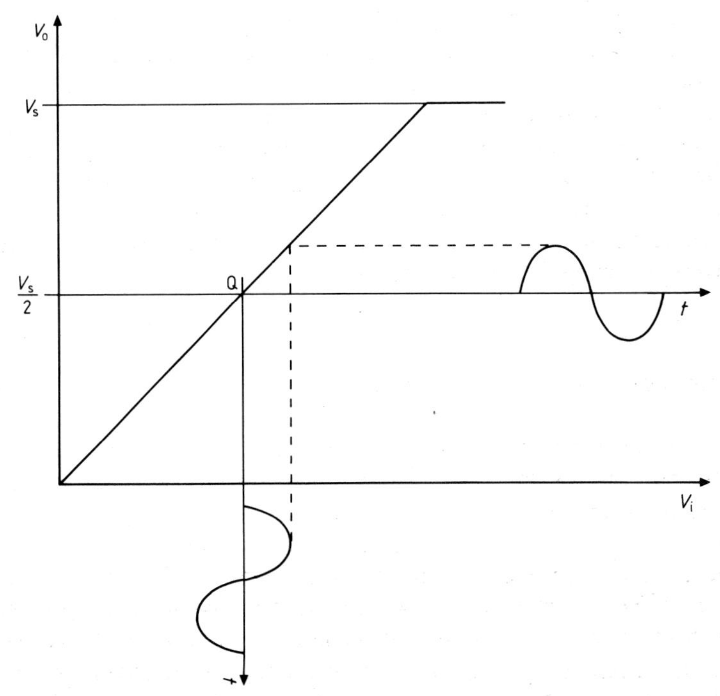

Figure 5.2 Idealised transfer characteristics for an emitter follower biased in the class A mode.

The instantaneous power drawn from the power supply is the product of the power supply current and voltage. V_s is constant and I_s, which is the same as I_e, varies sinusoidally about its mean value of $V_s/2R_e$. A sine wave has an average value of zero, so the average value of I_s is the same as the quiescent value. Hence the supply power is given by

$$P_s = (V_s/2R_e)V_s = V_s^2/2R_e. \qquad (5.2)$$

So the efficiency of power conversion, the ratio of P_o to P_s, is ¼ or 25%.

This mode of operation, in which current flows for the whole of a waveform period, is called class A. An efficiency of 25% is commonly quoted for a class A power amplifier, but three points should be noted.

(i) The result disregards bias arrangements and assumes idealised device properties. Practical circuits will have reduced efficiencies.

(ii) The figure has been obtained for a sinusoidal waveform. Other waveforms give, in general, different values. Many of the signals which need power amplification have waveforms with high peak to mean ratios. Most of the time their amplitude is small and they have only occasional excursions to large values. So the transistor spends most of the time developing little signal power while continuing to dissipate steady state power. Efficiencies with such waveforms can be very low, perhaps below 1%.

(iii) We have calculated the signal power developed in the internal load R_e. If power is required in an external load, capacitively or otherwise coupled to the amplifier, the efficiency may be lower.

5.2 Class B and C Operation

The efficiency in class A is low because both the transistor and the load resistor pass current and dissipate power at all times, even when there is no signal. An obvious way to reduce this dissipation and improve the efficiency is to shift the quiescent operating point so that the transistor does not conduct in the absence of a signal. In class B operation the transistor is biased at the point of cut-off as shown in figure 5.3, so that in our emitter follower example, $V_e = 0$. Now $I_s = I_e = 0$ and there is no quiescent power dissipation.

This reduction is bought at the expense of severe distortion of the signal waveform. Only positive excursions of the input signal result in an output signal, and negative excursions are completely suppressed. The output waveform for a sinusoidal input signal is thus a train of half sinusoids. However, it is possible to overcome this problem by adding another transistor to amplify the negative part of the signal. This second device could

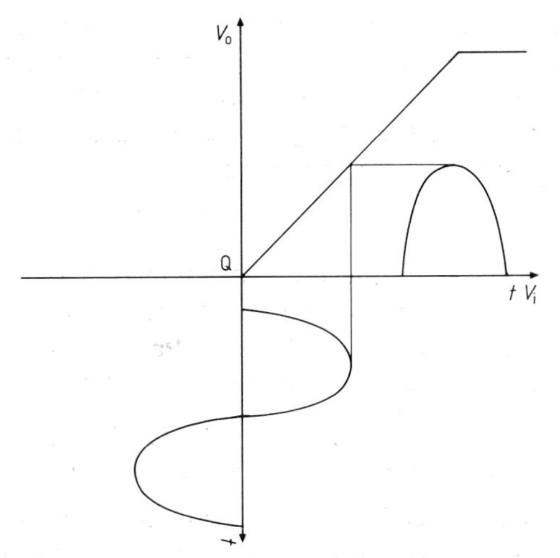

Figure 5.3 Class B operation of an emitter follower.

be another npn transistor, but it is very convenient if the reversed polarity property of a pnp device is exploited.

The two complementary devices are connected in a push–pull emitter follower configuration as shown in figure 5.4. They share a common load resistance R_e. A negative power supply is required for the pnp transistor and the circuit is symmetrical about the common rail. The same input signal is applied to the bases of both TR_1 and TR_2. When the input signal goes positive, TR_2 remains off but TR_1 conducts and the configuration behaves as a normal npn emitter follower with the output following the input. For negative signals, the role of the transistors is reversed. TR_1 is off but TR_2 conducts so that the output still follows the input.

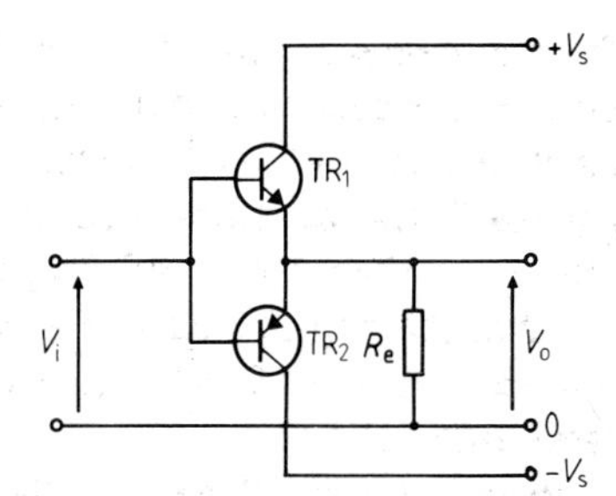

Figure 5.4 Complementary class B push–pull emitter follower.

The efficiency of the class B amplifier can be calculated in a similar fashion to the class A stage. For the maximum sinusoidal output of $2V_s$ peak to peak with idealised devices, the signal power delivered to the load is

$$P_o = (2V_s/2\sqrt{2})^2/R_e = V_s^2/2R_e. \qquad (5.3)$$

The waveform of the current drawn by the npn transistor from the positive power supply will be half sinusoidal with peak value of V_s/R_e as shown in figure 5.5. Averaging over one complete cycle leads to an expression for the mean current of

$$I_m = \frac{1}{2\pi}\frac{V_s}{R_e}\int_0^{2\pi} \sin\theta \, d\theta = \frac{V_s}{\pi R_e}. \tag{5.4}$$

The average current for the pnp transistor is the same, and so the total power supplied to the amplifier stage is

$$P_s = 2V_s^2/\pi R_e. \tag{5.5}$$

and the efficiency is $\pi/4$ or a little over 78%.

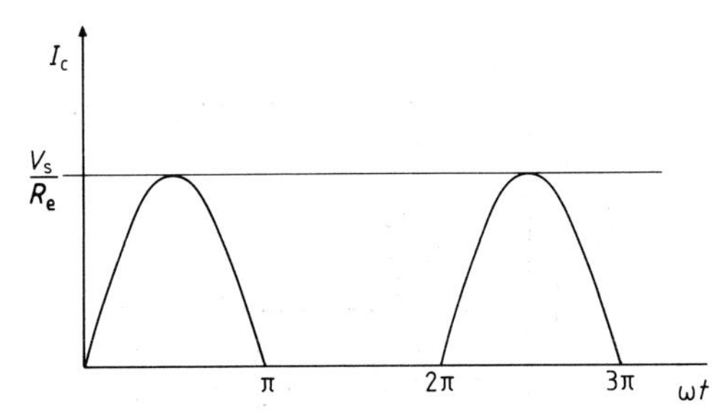

Figure 5.5 Half sinusoidal current waveform in a class B power amplifier.

This is a considerable improvement over the 25% for a class A stage. For signals which have a large ratio of peak to mean amplitude the efficiency remains high, and the advantages over class A are even more marked.

Practical class B amplifiers incorporate several additions to the basic circuit of figure 5.4. Because the power stage is an emitter follower with a voltage gain of unity, the input voltage swing must be large. An amplifier stage is needed to provide voltage gain. Also, BPT need about 0.6 V between base and emitter before they begin to conduct. So the circuit of figure 5.4 will have a dead zone for input signals between about −0.6 V and +0.6 V where both devices remain off. The output waveform will have crossover distortion as illustrated in figure 5.6.

Figure 5.7 shows the addition of a class A driver to the class B power stage which provides voltage gain and includes compensation for the forward bias required by the output transistor pair. The voltage drop across the two pn junction diodes provides the $2V_{be}$ needed to bias the output transistors at the turn-on point. Further refinements, including biasing for the class A stage and coupling capacitors, would be needed to complete a practical design.

Class C operation improves the efficiency further by shifting the bias point well into the cut-off region. Now conduction occurs only for brief

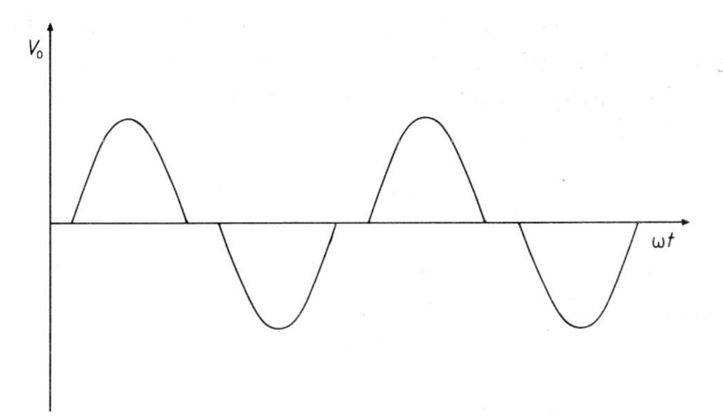

Figure 5.6 Output waveform from a class B amplifier with crossover distortion.

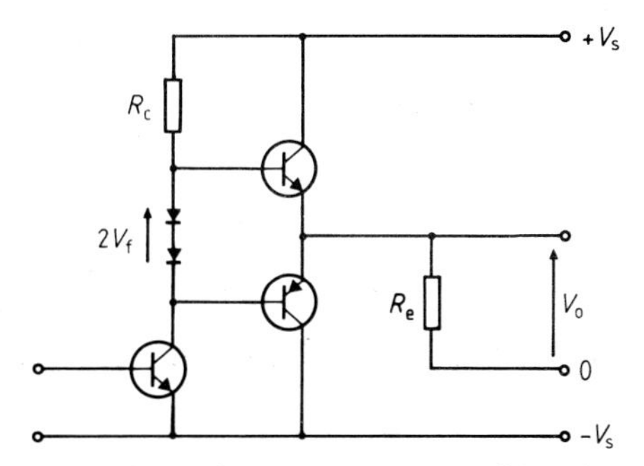

Figure 5.7 Complementary class B output stage with a class A driver (bias arrangements omitted).

periods at the peak of the input signal excursion and the current waveform is a series of short spikes as shown in figure 5.8. If the proportion of the cycle for which the transistor conducts is very small, the efficiency can approach 100%. A sinusoidal output can be recovered by a filter, usually an *LC* (inductor–capacitor) tuned circuit, which can select the fundamental or a harmonic from the periodic waveform and reject unwanted frequency components.

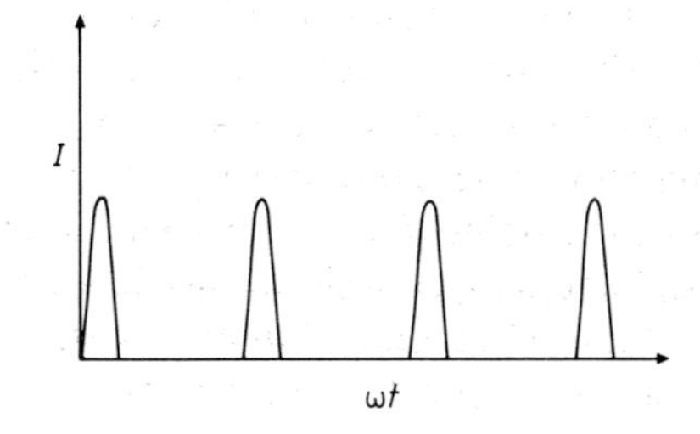

Figure 5.8 Current waveform in a class C power amplifier.

5.3 Power Supplies

The availability of very low power consumption integrated circuits means that an increasing number of electronic circuits can be powered by batteries. Either primary or rechargable secondary cells can be used and both types give the advantages of portability and isolation from the mains supply. However, there are still many circuits, especially those which produce large output powers, that need to be operated from the mains supply.

A transformer is normally used to reduce the mains voltage (240 V at 50 Hz in the UK) to a lower level and to provide isolation between the powered circuit and the mains supply. The output from the transformer can then be made unidirectional (rectified) with one or more diodes. A simple rectifier circuit using a single diode is shown in figure 5.9(*a*) where R_l is the load resistance representing the circuit attached to the power supply. The transformer delivers an approximately sinusoidal output of V_p peak.

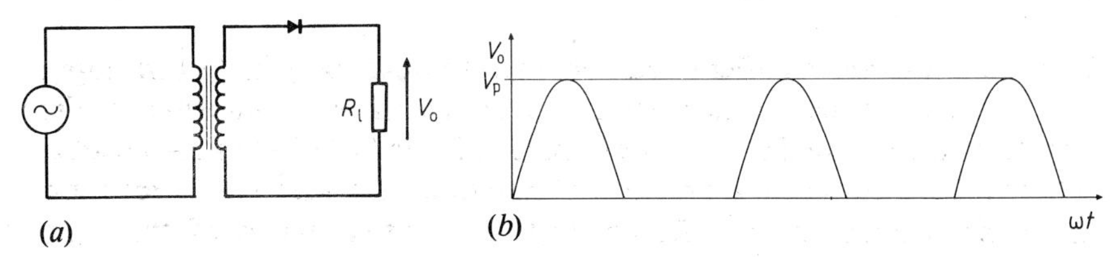

Figure 5.9(*a*) Half-wave rectifier driven from a stepdown transformer (*b*) output voltage waveform for a half-wave rectifier.

On the positive half cycles of the transformer output, the diode conducts and current flows through the load developing an output voltage which rises to V_p if the forward voltage drop across the diode is neglected. Figure 5.9(*b*) shows the waveform developed across the load. Although unidirectional, it is far from steady. A more constant voltage can be achieved by adding a capacitor across the rectifier output, as shown in figure 5.10(*a*), to store charge during the time that the diode conducts and to supply output current into the load during the remainder of the cycle.

The output waveform can be regarded as a constant level V_o with a superimposed ripple V_r as shown in figure 5.10(*b*). If the capacitor is sufficiently large, the voltage does not fall substantially from the peak value V_p during the discharge part of the cycle. So the load current I_l remains approximately constant. Also the charge time t_c will be small compared with the mains period t and the discharge time $(t - t_c)$ can be taken as t. Then the peak to peak ripple voltage V_r is the drop in output voltage during the discharge time and is given by

$$V_r = I_l t/C \quad (5.6)$$

$$= V_p t/R_l C. \quad (5.7)$$

The ripple waveform is approximately symmetrical about V_o, and so the output voltage can be written as

$$V_o = V_p - V_r/2. \quad (5.8)$$

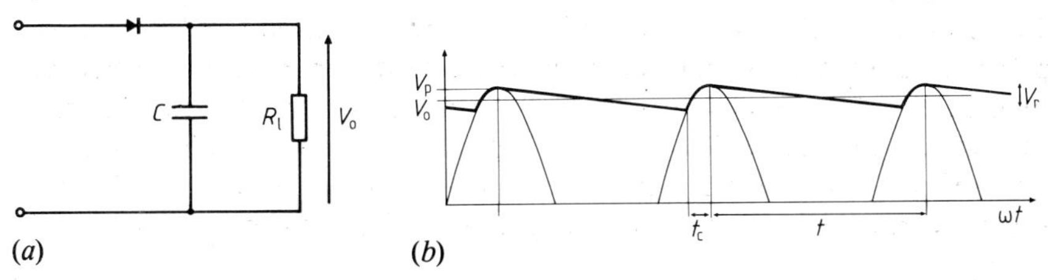

Figure 5.10(*a*) Halfwave rectifier with a reservoir capacitor (*b*) the resultant smoothed output waveform with a ripple voltage V_r superimposed on the mean output V_o.

This is a half-wave rectifier circuit because it utilises only half of the input waveform. Better use of the transformer can be made if both halves of the waveform are rectified in a full-wave circuit. Figure 5.11 shows one possible full-wave rectifier using a bridge configuration of diodes with a reservoir capacitor. Diodes D_1 and D_3 conduct on positive halves of the input waveform while D_2 and D_4 are reverse biased. On the negative half cycles D_1 and D_3 are non-conducting and current flows through D_2 and D_4. In both cases current flows through the load in the same direction. The results in equations (5.6)–(5.8) apply except that the discharge time t becomes half of the mains period.

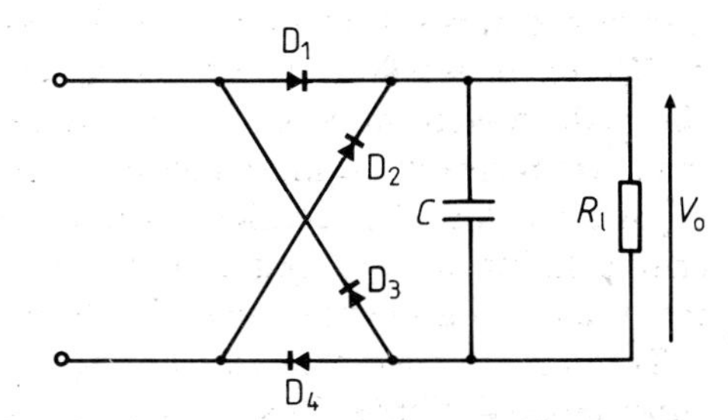

Figure 5.11 Full-wave rectifier with a reservoir capacitor.

In principle, the ripple output from a half-wave or a full-wave rectifier can be made as small as desired by increasing the value of the reservoir capacitor as indicated by equations (5.6) and (5.7). However, there are practical and economic reasons why very large capacitors cannot be used. As the value of the capacitance is increased, the time available for charging is reduced. Hence the magnitude of the charging current pulse must be increased. Eventually the ratings of the transformer, diodes and capacitor will be exceeded. Moreover, large capacitors are also expensive.

5.4 Regulation and Stabilisation

The quality of the output from a simple rectifier circuit can be improved by stabilising or regulating the output voltage. Proper regulation means maintaining the output voltage constant while the load current (or load resistance) is varied. Stabilisation implies keeping the output constant irrespective of the cause of any change and includes regulation. However, the two terms are often considered to be interchangable.

A Zener diode provides a simple way of stabilising a voltage output as illustrated in figure 5.12. The diode is operated in the reverse breakdown region shown in figure 2.6 where the reverse voltage becomes almost independent of the current. V_i must be greater than V_z and the series resistance R_s absorbs the difference. The value required for R_s is easily determined: if the load current is $I_l = V_z/R_l$, and the Zener current is I_z, the total current through R_s is $I_l + I_z$. Then Ohm's law gives

$$R_s = (V_i - V_z)/(I_l + I_z). \tag{5.9}$$

In practice, the application of equation (5.9) is complicated by tolerances in the components and by the possible variations in I_l. It is usually necessary to allow a large voltage drop across R_s to ensure satisfactory operation.

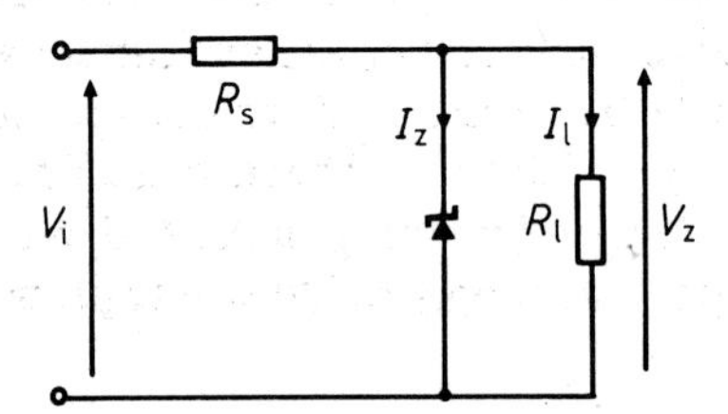

Figure 5.12 Simple Zener diode voltage stabilisation circuit.

The stability of the output voltage from this circuit depends on the slope of the Zener diode characteristic in the breakdown region. If the deviation from the steady state operating point is small, the behaviour can be approximated by a straight line and described by a small signal dynamic slope resistance $r_z = \mathrm{d}V_z/\mathrm{d}I_z$. Practical Zener diodes have dynamic resistances between about 1 and 100 Ω.

Variations in the load current I_l must be absorbed by the Zener diode and produce changes in I_z according to equation (5.9). Fluctuations in the input voltage V_i arising from ripple or from variations in the mains supply voltage also lead to changes in I_z from equation (5.9). All of these changes superimposed on the steady state values can be treated as signals.

A small signal model for the stabilising circuit of figure 5.12 can then be derived as shown in figure 5.13. The signals v_i, v_z, i_l and i_z represent changes in V_i, V_z, I_l and I_z respectively. The model for the Zener diode is just the dynamic resistance r_z. If the effects of v_i and i_l are considered separately, the output signal which results from an input voltage variation v_i can be written as

$$v_z = \frac{r_z}{(R_s + r_z)} v_i \tag{5.10}$$

because R_s and r_z form a potential divider. The change in output which results from variations in load current is given by

$$v_z = i_l r_o \tag{5.11}$$

where r_o is the output resistance of the stabilising circuit. Application of the definition of output resistance quoted in §4.6 shows that r_o is the parallel combination of R_s and r_z, which will be only a little less than r_z because R_s is normally very much greater than r_z.

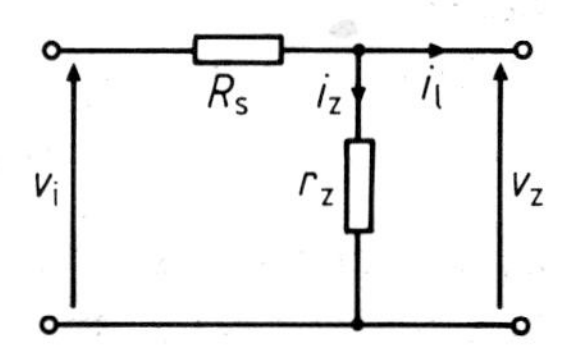

Figure 5.13 Model of the Zener diode stabilisation circuit for ripple voltage calculation.

This circuit is adequate for low load currents. If high output currents are needed, an emitter follower can be added to increase the available current as indicated in figure 5.14. Here V_z supplies the base voltage of a transistor. The output at the emitter is $V_z - V_{be}$ and is fairly well defined. The current gain of the transistor reduces the demand on the Zener reference circuit to I_l/β. Note that the collector is returned to the raw positive supply rail to ensure proper biasing of the collector–base junction.

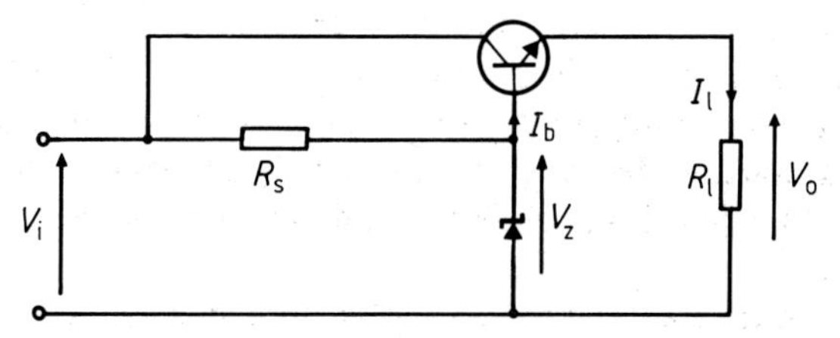

Figure 5.14 Emitter follower assisted Zener diode.

More sophisticated regulator circuits can use operational amplifiers with feedback to compare the output voltage with a reference and apply appropriate correction. An outline circuit using a transistor to augment the current output from an operational amplifier is shown in figure 5.15. The differential input to the operational amplifier must be small and hence the output voltage must be

$$V_o = (1 + R_1/R_2)V_z. \tag{5.12}$$

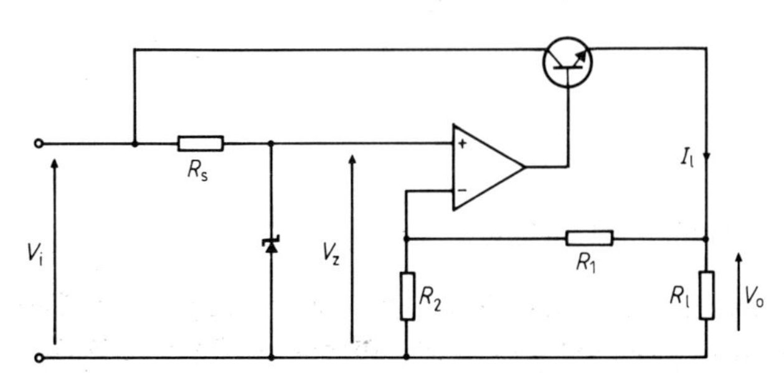

Figure 5.15 Voltage stabiliser using an operational amplifier.

Note that the base–emitter voltage of the transistor does not enter into equation (5.12) and the stability of this circuit is determined almost entirely by the quality of the reference voltage. Simple Zener diodes may be inadequate for the highest performance and precision reference circuits with accurate output voltages which incorporate compensation for temperature variation may be needed. More elaborate circuits include adjustable output voltage and may have current limiting to prevent damage from overload. Regulators in IC form with built-in voltage references, error amplifiers and power transistors are readily available in a variety of fixed standard voltages and also with adjustable outputs.

6

Signal Conditioning

6.1 *RC* Networks and Filtering

All useful signals contain a variety of frequency components. Filtering is an operation which alters the relative proportions of the components, reducing some and increasing others. In an extreme case some signal components may be removed entirely. Simple filtering can be achieved just with combinations of resistors and capacitors, but more complicated filters may need active devices or other circuit elements.

The *RC* network shown in figure 6.1 is a low pass filter, transmitting steady state quantities and low frequency signals freely, but attenuating high frequency components. It is widely used to restrict the bandwidth of a signal or to reduce fluctuations in a nominally steady voltage or current. The resistance and capacitance form a potential divider whose division ratio is a function of frequency. At very low frequencies the capacitor reactance $1/\omega C$ becomes very high, so the low frequency signal loss through the divider is negligible. At very high frequencies the reactance becomes very small and signal is heavily attenuated.

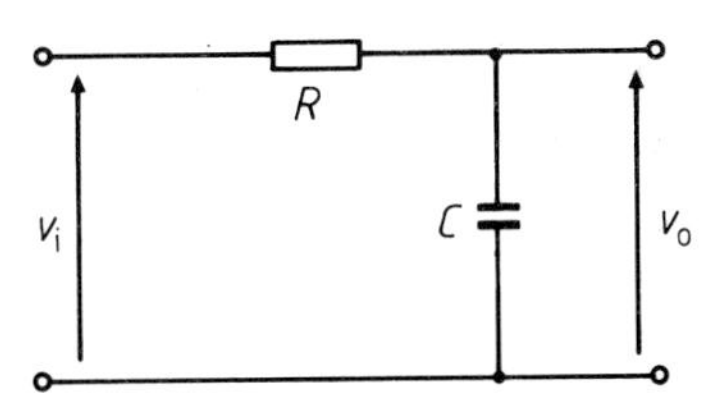

Figure 6.1 *RC* low pass filter.

For many purposes it is sufficient to say that the transition between transmission and attenuation occurs at a frequency where the capacitor reactance is comparable with the resistance. It is mathematically convenient to identify a transition or corner frequency f_c defined by equating the resistance to the magnitude of the capacitor reactance. So $R = 1/\omega_c C$ or $\omega_c CR = 1$, where ω_c is the angular frequency $2\pi f_c$. Rather approximately we can say that all signal frequency components below f_c are transmitted and all above rejected.

More detailed analysis of the behaviour leads to the result shown in figure 6.2. The transmission v_o/v_i is plotted against frequency using logarithmic scales. For very low frequencies, $v_o/v_i = 1$. At f_c the magnitude of the response is 3 dB below the low frequency level. Well above the corner frequency the response is inversely proportional to the frequency which corresponds to a reduction of 20 dB (a factor of 10) for every decade increase in frequency. It is also equivalent to a fall of approximately 6 dB per octave, and both terms are in common use.

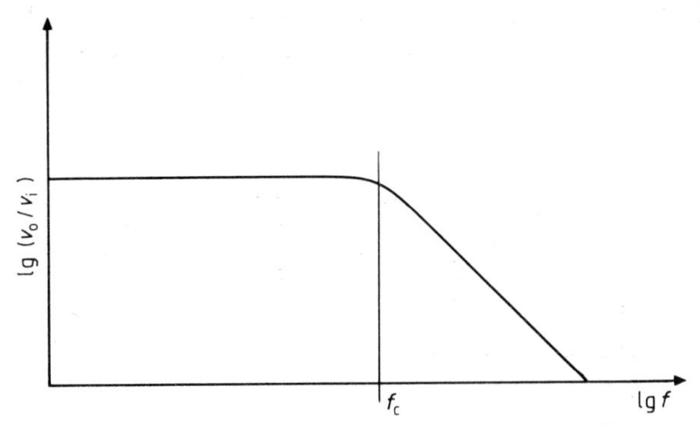

Figure 6.2 Frequency response of a low pass filter.

The *CR* network shown in figure 6.3 is really the same as the *RC* network except that the output signal is taken across the resistive branch of the divider. The frequency response will be the inverse of the *RC* network, blocking steady state (zero frequency) quantities, attenuating low frequency signals and transmitting high frequency components. The corner frequency of this high pass filter is again defined by $\omega_c CR = 1$. Figure 6.4 shows the form of the frequency response.

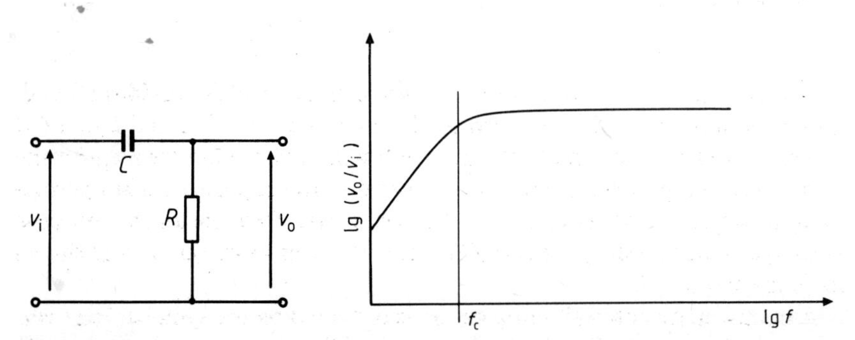

Figure 6.3 *CR* high pass filter. **Figure 6.4** Frequency response of a high pass filter.

It is important to realise that although the *RC* and *CR* networks are usually presented as shown in figures 6.1 and 6.3, they may often be disguised in real circuits. We have already met examples of this. The input capacitor of the single stage amplifier in figure 4.6 forms part of a *CR* network, but the resistance is not just the resistor R_2. Inspection of figure 4.7 shows that it is the effective input resistance, the parallel combination of R_1, R_2 and R_{be}.

If we require sharper cut-off than the 20 dB per decade provided by a simple *RC* or *CR* filter, multiple stages may be used to improve the performance. Figure 6.5(*a*) shows two *RC* stages cascaded to provide a fall-off of 40 dB per decade at high frequencies. Unfortunately the stages interact and the corner frequency is no longer given by $\omega_c CR = 1$. Interposing a buffer stage as shown in figure 6.5(*b*) isolates the filters and avoids the problem.

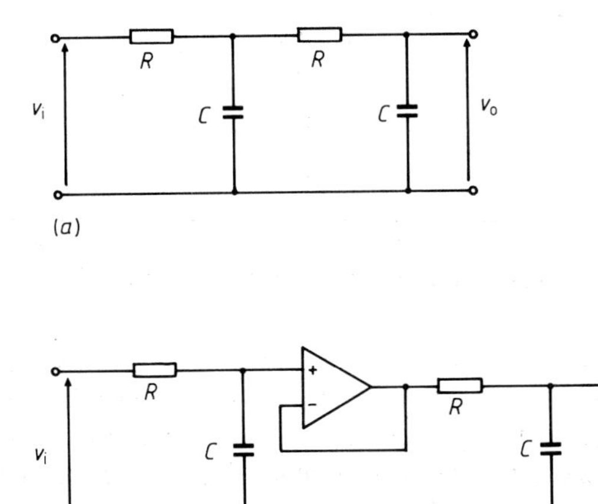

Figure 6.5 Two *RC* stages (*a*) directly cascaded, (*b*) separated by a voltage follower buffer.

The symmetrical Wien network shown in figure 6.6(*a*) provides a band-pass function. It can be regarded as the combination of an *RC* and a *CR* filter. It is easy to see that both at very low and very high frequencies the signal transmission must tend to zero because the capacitors will approximate to open and short circuits. At intermediate frequencies the response rises to a rather broad peak at $\omega_p CR = 1$, with a transmission of ⅓ as shown in figure 6.6(*b*).

A bandstop or notch filtering action is provided by the symmetrical twin-tee network shown in figure 6.7(*a*). Again it will be obvious that at extremely

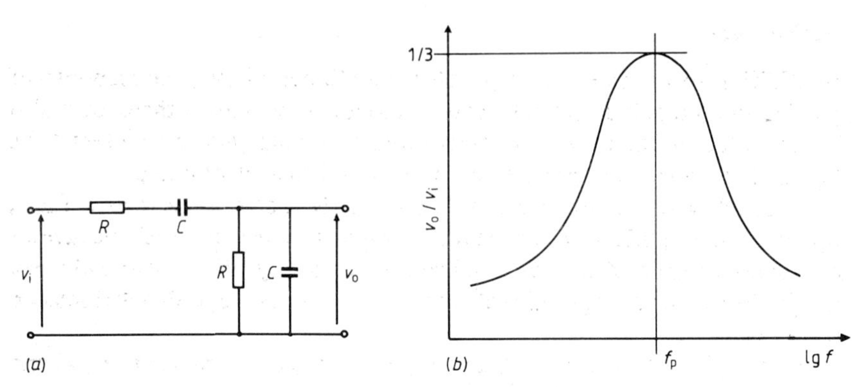

Figure 6.6 Frequency response of a Wien network.

low and high frequencies the capacitors will appear as open and short circuits and hence there will be free transmission of signals. In between these extremes the response falls to zero at a frequency defined by $\omega_z CR = 1$ as shown in figure 6.7(*b*). In practice the network components are unlikely to be matched exactly, and some transmission does occur. Nevertheless the notch can still be sharp.

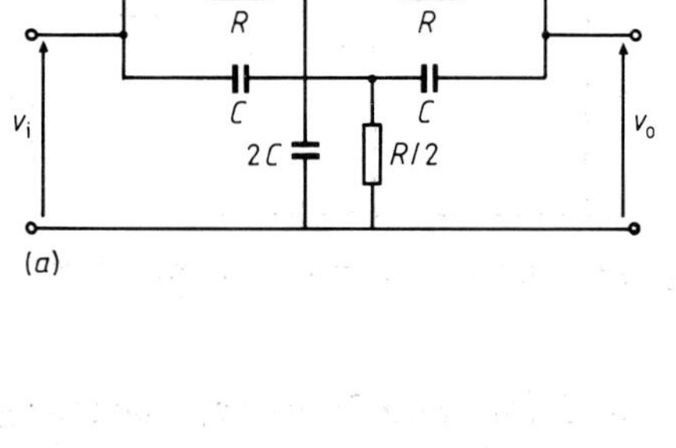

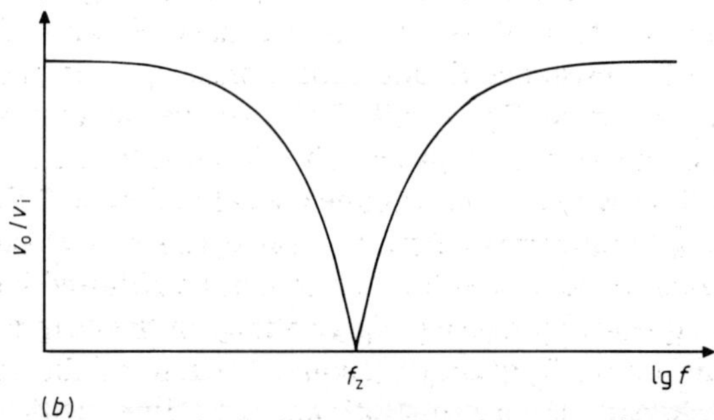

Figure 6.7 (*a*) Twin-tee network, (*b*) its frequency response.

Active filters

In addition to the interaction problem mentioned above, the response of passive *RC* filters is affected by external loading. Moreover there must also be signal loss in the networks. Active filters provide gain to counteract the loss and can give more precise and more sophisticated filtering.

Although any amplifier can be used as the basis of an active filter, operational amplifiers are particularly convenient because their behaviour can be precisely defined by external components only. We have already seen this in the case of the operational amplifier with resistive feedback discussed in §4.10.

If we replace the resistors R_i and R_f by more general impedances, Z_i and Z_f, we get the circuit of figure 6.8. Exactly the same analysis can then be applied while working with signal quantities. The gain $A_v = v_o/v_i$ will be

$$A_v = -Z_f/Z_i. \tag{6.1}$$

This result reduces to equation (4.37) if Z_i and Z_f are purely resistive, in which case there is no frequency dependence apart from limitations in the operational amplifier.

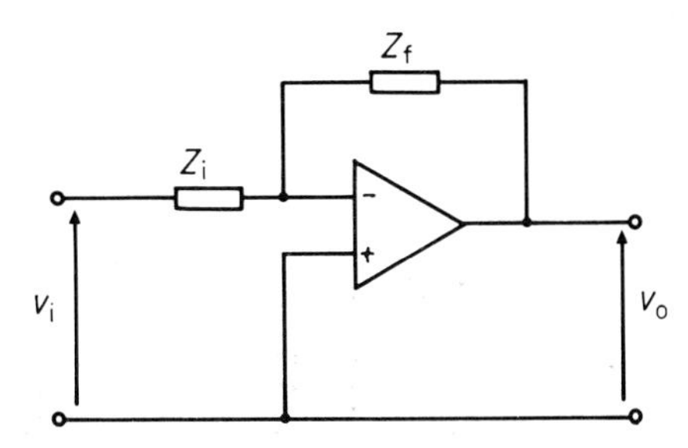

Figure 6.8 Active filter using an operational amplifier.

However, if Z_i or Z_f vary with frequency, then the gain A_v will also. For example, suppose that Z_i is just a resistance R_i and Z_f is a parallel combination of a resistance R_f and capacitance C_f. Z_f is then a function of frequency as shown in figure 6.9. This will be directly reflected in the frequency response of the amplifier. The circuit is then an active version of the low pass filter with a corner frequency defined by $\omega_c R_f C_f = 1$.

At very low frequencies where the capacitance is effectively an open circuit, the gain reduces to $-R_f/R_i$, and can be adjusted independently of the corner frequency by altering R_i. Loading on the output will have little effect on gain or on frequency response because of the very low output resistance inherent in an operational amplifier with this feedback configuration.

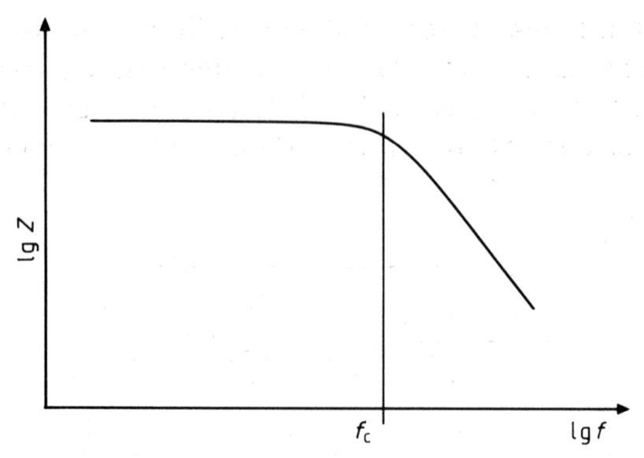

Figure 6.9 Variation of Z_f with frequency.

Replacing R_i by a series combination R_i and C_i as shown in figure 6.10 gives a bandpass filter. The input network provides a high pass function with a corner frequency ω_u defined by $\omega_u C_i R_i = 1$. Low pass filtering comes from the feedback network which has a corner frequency ω_l defined by $\omega_l C_f R_f = 1$. The two networks are isolated by the virtual earth at the inverting input and do not interact, so the overall frequency response is as shown in figure 6.11.

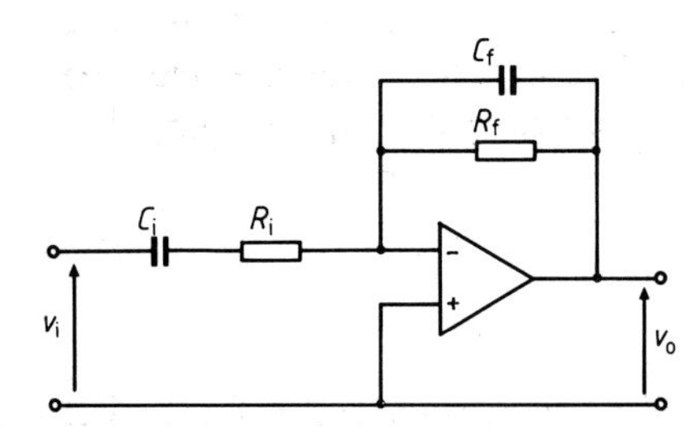

Figure 6.10 Active bandpass filter.

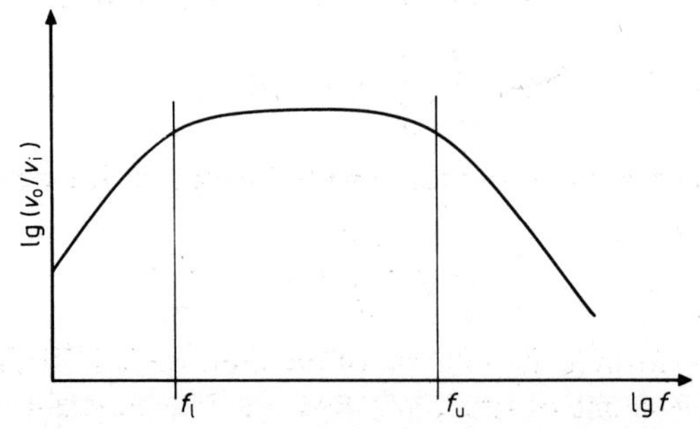

Figure 6.11 Frequency response of a bandpass filter.

There is no need to restrict the input and feedback loops to two-terminal networks. Three-terminal networks such as the twin-tee can be used as shown in the example of figure 6.12. It is fairly easy to deduce the form of the frequency response of this filter, although an exact analysis is more difficult.

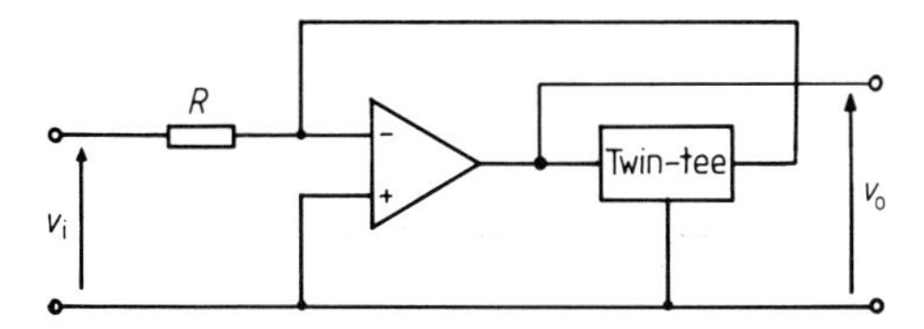

Figure 6.12 Narrow bandpass filter using a twin-tee network and feedback.

The virtual earth argument tells us that there will be a constant current of v_i/R_i driven through R_i from the input voltage source. This must be balanced by a current through the twin-tee in the feedback loop. Although the twin-tee network in this circuit has different termination conditions from the example in figure 6.6 (low resistance on both sides), it still exhibits a notch response. At very high and very low frequencies there is considerable feedback and the circuit voltage gain is small. In the middle range the attenuation of the twin-tee increases, so the output voltage must rise to maintain the feedback current. At the notch frequency the output will reach a sharp peak. The circuit has a bandpass response which is similar to, though not of exactly the same form as, a tuned circuit made with inductance and capacitance.

6.2 Integration and Differentiation

Integration and differentiation are special cases of filtering. In figure 6.13 the current i into the capacitor is $i = \mathrm{d}q/\mathrm{d}t$ and the voltage across the capacitor is $v_o = q/C$. So $v_o = (1/C)\int i\,\mathrm{d}t$. But $i = (v_i - v_o)/R$. Then

$$v_o = \frac{1}{RC}\int (v_i - v_o)\,\mathrm{d}t. \tag{6.2}$$

If the output voltage is small compared with the input voltage this reduces to

$$v_o = \frac{1}{RC}\int v_i\,\mathrm{d}t. \tag{6.3}$$

So the output waveform is the integral of the input signal. The product RC, which has the dimensions of time, is called the time constant of the circuit and is often denoted by the symbol τ.

The requirement of the output voltage being small compared with the input for the approximation to hold corresponds to keeping the time scale short compared with τ. So the *RC* network only integrates accurately for frequencies well above the corner frequency $\omega_c = 1/RC = 1/\tau$. We see that integration is equivalent to a frequency response continuously falling at a rate of 20 dB per decade.

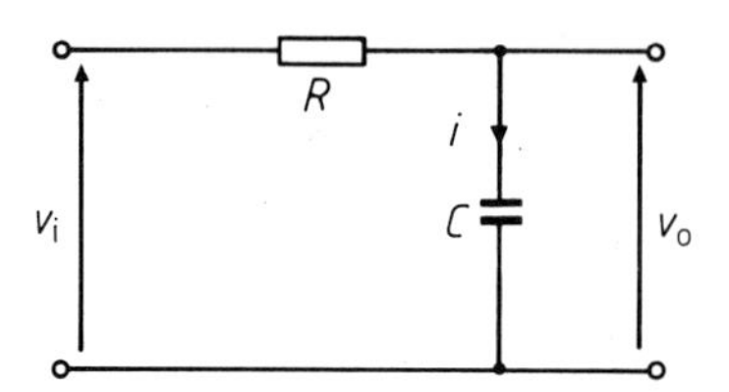

Figure 6.13 *RC* integrating network.

A similar analysis shows that the output from a *CR* network can be the differential of the input waveform.

$$v_o = RC\,\mathrm{d}v_i/\mathrm{d}t. \tag{6.4}$$

Again there is a restriction on the magnitude of the output voltage if the operation is to be precise. In this case the requirement corresponds to the time scale being long compared with the time constant. So differentiation is only accurate for frequencies well below the corner frequency, in the region where the response is rising at 20 dB per decade.

Although passive *RC* and *CR* networks can only give accurate integration and differentiation over a limited frequency range, they are widely used outside these ranges. A common use of the *CR* network is to provide short pulses synchronised to the edges of a train of rectangular pulses. Figure 6.14 shows the pulse shaping effect of this circuit, which is still referred to as a differentiating network although it is really the high pass filtering property that is being exploited in this application. In a similar way the *RC* low pass filter is often referred to as an integrating network even when its function is that of bandwidth restriction or averaging by the reduction of high frequency signal components.

If high precision integration or differentiation is required and we cannot accept the restriction on the output signal amplitude used in the derivation of equations (6.3) and (6.4), the operational amplifier can provide a solution. Figure 6.15 shows an operational amplifier connected in the inverting configuration with capacitive feedback and a resistive input. Using the virtual earth approximation we can immediately write down equations for the input and output loops

$$v_i = iR \qquad v_o = \frac{-1}{C}\int i\,\mathrm{d}t$$

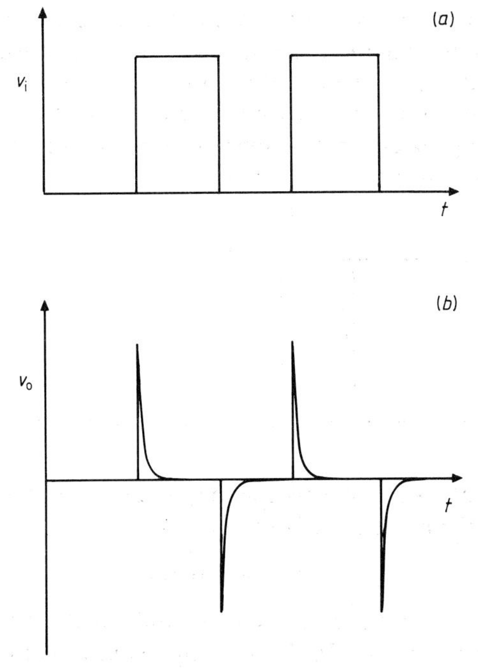

Figure 6.14(*a*) Input, and (*b*) output waveforms for a *CR* differentiating network.

and hence

$$v_o = \frac{-1}{RC} \int v_i \, dt. \tag{6.5}$$

The presence of the virtual earth at the inverted input isolates the capacitor from the input loop so that the input current, and hence the capacitor charging current, does not depend on the output voltage. The restrictions necessary in the passive integrator are removed, and apart from the limitations inherent in the virtual earth approximation and other operational amplifier defects, the integration is perfect.

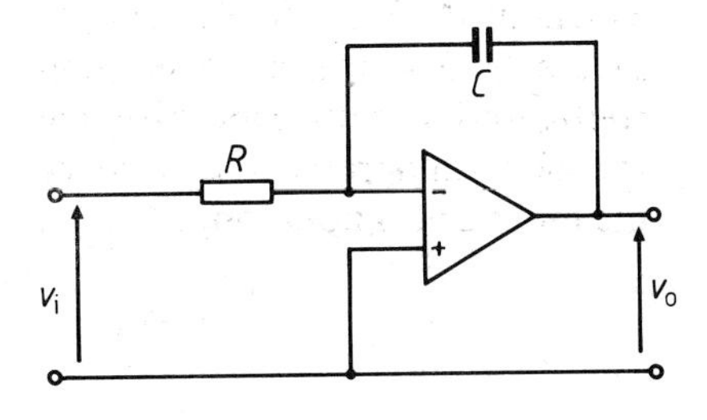

Figure 6.15 An active integration circuit.

It should be obvious that an active differentiating circuit can be made with an operational amplifier and a *CR* network. The virtual earth approximation leads to the result

$$v_o = -RC\,dv_i/dt. \tag{6.6}$$

However, perfect differentiation implies a continuously increasing output with increasing frequency, whereas general purpose operational amplifiers have a low pass response. So the virtual earth approximation soon fails. Also differentiators tend to accentuate noise on the input signal. For these reasons the practical applications of active differentiators are limited.

Summation

Two or more signals can be approximately summed by a simple resistive network as shown in figure 6.16. The input currents are

$$i_1 = (v_1 - v_o)/R_1 \qquad i_2 = (v_2 - v_o)/R_2$$

and the output voltage v_o must be given by

$$v_o = iR = (i_1 + i_2)R.$$

If v_o is negligible in comparison with v_1 and v_2

$$v_o = \left(\frac{v_1}{R_1} + \frac{v_2}{R_2}\right)R. \tag{6.7}$$

So the output is the sum of the input signals scaled by the factors $1/R_1$ and $1/R_2$.

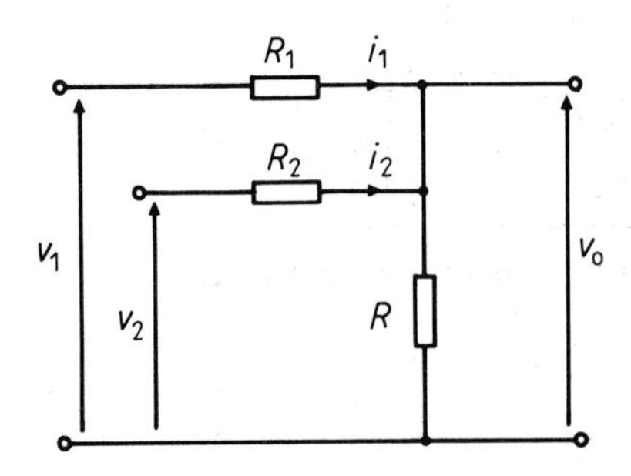

Figure 6.16 Summation of two input signals in a resistive network.

Again an active version, using an operational amplifier and invoking the virtual earth approximation, will remove the restriction on the magnitude of the output voltage. Figure 6.17 shows an active summation circuit with three inputs. The output is easily shown to be exactly

$$v_o = -R\left(\frac{v_1}{R_1} + \frac{v_2}{R_2} + \frac{v_3}{R_3}\right) \tag{6.8}$$

provided that the virtual earth approximation is valid.

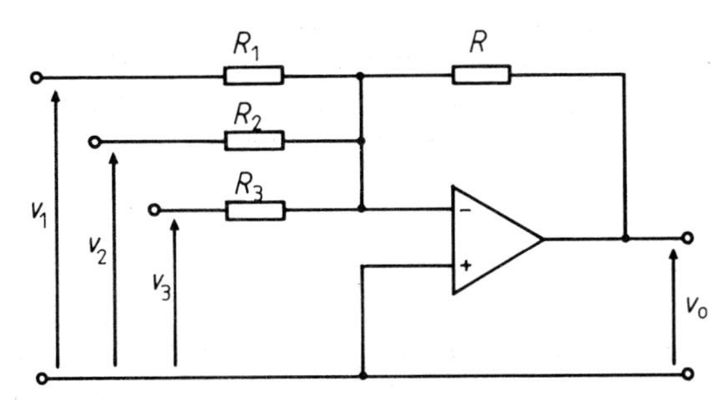

Figure 6.17 Active summation of three input signals.

More than one of these operations can be combined in a single circuit. Figure 6.18 shows an example where an operational amplifier is used to produce the integral of the sum of two input signals. Application of the virtual earth argument quickly yields an expression for the output voltage

$$v_o = \frac{-1}{C}\int\left(\frac{v_1}{R_1} + \frac{v_2}{R_2}\right) dt. \tag{6.9}$$

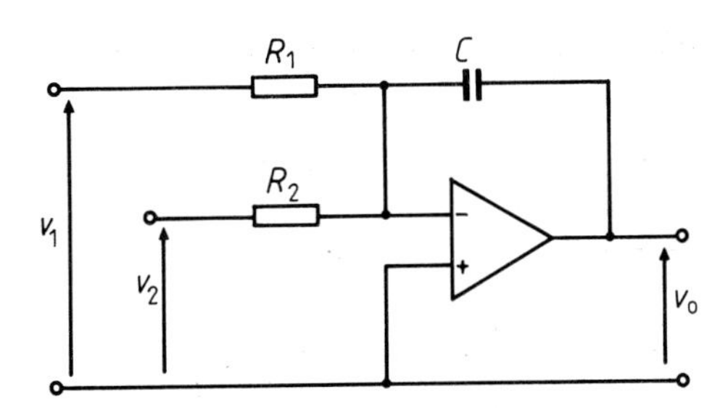

Figure 6.18 Active integration of the sum of two signals.

6.3 Non-linear Processing

In linear signal processing all of the devices and components in the processing chain are operated in their linear regions. The amplitude of the output is directly proportional to the amplitude of the input signal. Non-linear operations can be accomplished by using the properties of a diode or similar device. One of the simplest non-linear circuits uses a diode and a resistor as shown in figure 6.19.

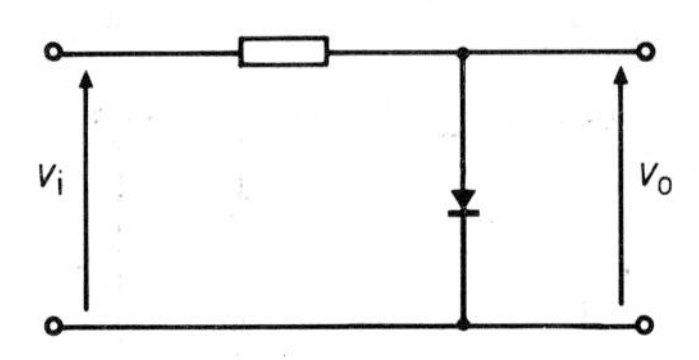

Figure 6.19 Non-linear circuit known as a resistor–diode limiter.

As with the *RC* and *CR* networks we shall assume that the loading on the output of the network is negligible. For positive values of the input signal v_i the diode is forward biased and conducts. So the output voltage v_o is limited to the forward voltage drop of the diode V_f. However, for negative input signals the diode is reverse biased and non-conducting. Then the input signal is transmitted directly to the output.

So this network discriminates between positive and negative signals, allowing negative excursions to pass unaltered, but limiting the positive amplitude to V_f. If the signal amplitude is large, it is common to neglect V_f and consider that the limiting is to zero. Figure 6.20 illustrates the limiting effect on a waveform.

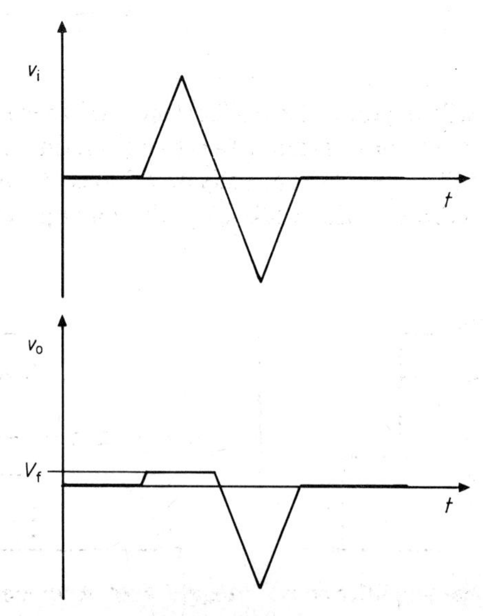

Figure 6.20 Positive limiting of a waveform.

More general limiting to any level can be achieved if the diode is returned to a reference voltage V_r as shown in figure 6.21. Now the positive signal excursion is limited to $(V_r + V_f)$ whilst all signal values below that level are still unaffected. It is obvious that these circuits can be made to limit negative input signal excursions by reversing the polarity of the diode.

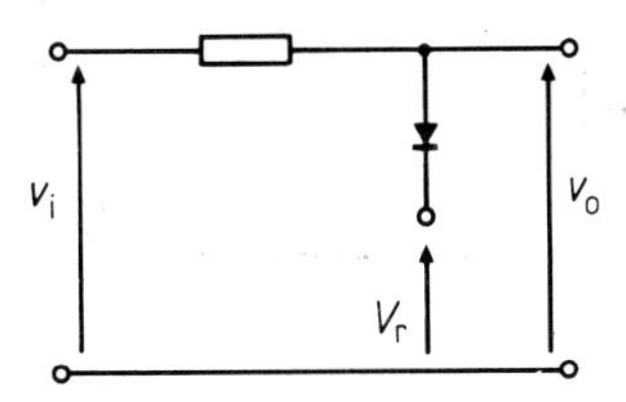

Figure 6.21 Limiting to a reference voltage V_r.

Interchanging the positions of the diode and resistor gives the network shown in figure 6.22. This is the unsmoothed half-wave rectifier of §5.3 which passes all positive signals but suppresses any negative inputs. Again a reversed polarity version is possible.

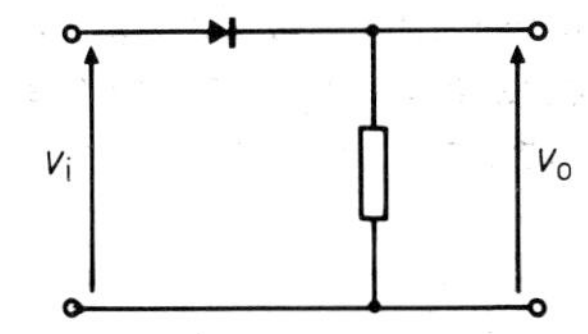

Figure 6.22 Diode–resistor polarity discriminator.

Loading effects will degrade the performance of these passive circuits. Active versions using operational amplifiers can overcome these problems as with the *RC* and *CR* filter networks already mentioned. Figure 6.23 shows an active limiter using a diode in the feedback path of an operational amplifier.

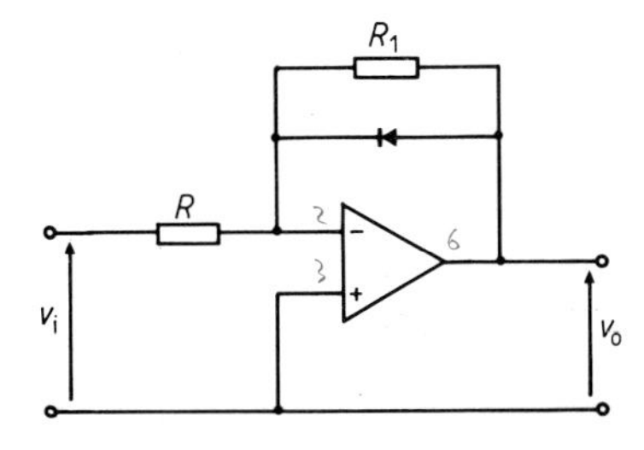

Figure 6.23 Operational amplifier active limiter.

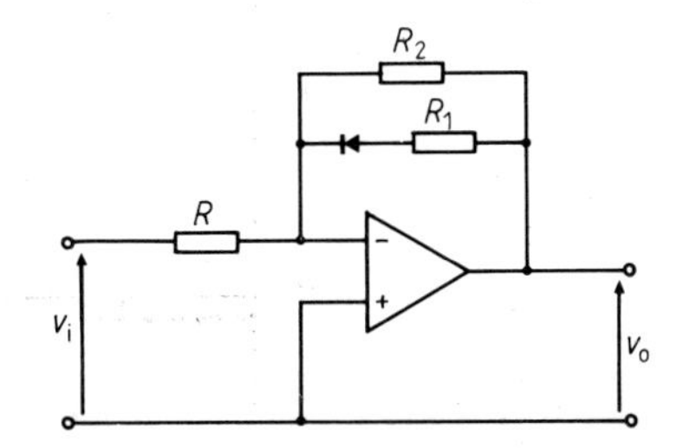

Figure 6.24 A diode–resistor network providing asymmetric voltage gain.

For negative input signals the output is driven positive and the diode conducts. The output is then limited to V_f. For positive inputs the diode is reverse biased and the output voltage is given by the usual expression for the inverter $-(R_1/R)v_i$ (equation (4.36)). An obvious variant of the circuit includes a resistor in series with the diode as shown in figure 6.24. Now we have a circuit which has a voltage gain proportional to R_1 for positive outputs and to the parallel combination of R_1 and R_2 for negative outputs.

6.4 Logarithmic Amplification

These circuits have taken a very simple two-region model for the behaviour of a diode. This model is quite adequate for describing the operation of limiters, where the signals are assumed large compared with V_f. But for small signals, of less than one volt in amplitude say, we need a more accurate description of the device properties.

In §2.2 we saw that the current voltage relation for a pn junction diode can be expressed by the exponential equation (2.1). If we use this expression to describe the behaviour of the diode in the circuit of figure 6.25, we shall get a more precise picture of the circuit operation.

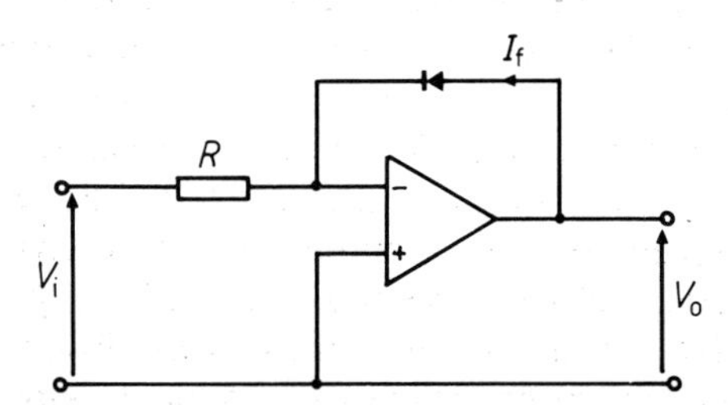

Figure 6.25 Logarithmic amplification with a diode.

It will be adequate to analyse this circuit for steady state voltages. Then we can derive an expression for the feedback current through the diode

$$I_f = I_s\left[\exp\left(\frac{qV_f}{\eta kT}\right) - 1\right] = -\frac{V_i}{R}. \tag{6.10}$$

For all reasonable values of the diode current I_f the exponential term is very much greater than unity. Approximating, rearranging and taking natural logarithms leads to an expression for the output voltage

$$V_o = V_f = \frac{\eta kT}{q}\ln\left(\frac{-V_i}{I_s R}\right). \tag{6.11}$$

So the output of this circuit is proportional to the logarithm of the input voltage. Practical diode circuits can show accurate logarithmic behaviour over two or three decades of current. If precise logarithmic operation is required over a wider range, the exponential relation between the collector current and the base–emitter voltage of a BPT can be used as shown in figure 6.26.

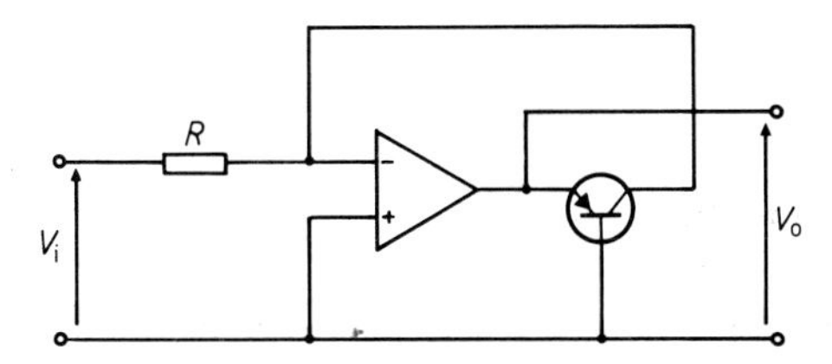

Figure 6.26 Logarithmic amplification using a transistor.

Equation (6.11) shows that the logarithmic relation will be temperature dependent because of the kT term and also because I_s is a sensitive function of temperature. Practical versions of diode and transistor logarithmic converters usually include extra devices and components to provide compensation for temperature variations.

Incorporating a pn junction diode in the input loop of an operational amplifier will give an input current that varies exponentially with the input voltage. Resistive feedback then translates the current to an output voltage so that the circuit provides an antilogarithmic response.

Combinations of logarithmic and antilogarithmic amplifiers and summing circuits can perform multiplication and division of analogue signals. The accuracy of such operations depends on precise control and matching of the device non-linearities, and on careful compensation for any temperature effects. Integrated circuits are available with these features.

6.5 The Comparator

If a high gain differential amplifier—perhaps an operational amplifier—is operated without negative feedback, even quite small input signals will be sufficient to generate large output voltages. However, the output excursions must be limited by saturation or cut-off of the transistors in the internal amplifier circuit.

A typical operational amplifier without feedback has a transfer characteristic as shown in figure 6.27. For differential input signals of less than a millivolt or so, the amplifier exhibits voltage gain and is operating in the linear region. Signals outside this range drive the amplifier output into saturation. In practice, the saturation levels are a little less than the voltages of the power supplies to the amplifier. For example, an operational amplifier operated from the standard +15 and −15 V power rails might saturate at +14 and −14 V.

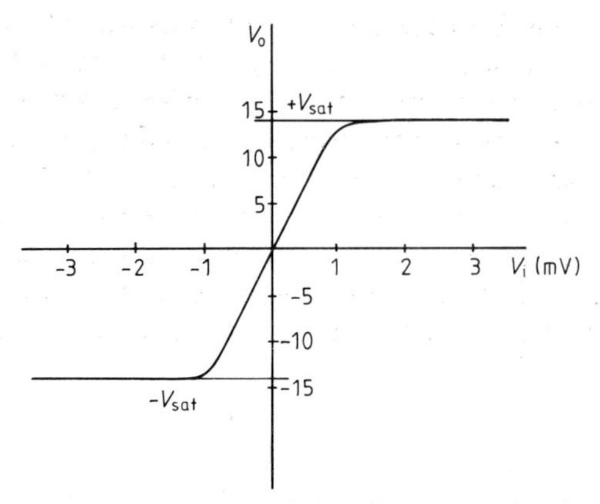

Figure 6.27 Transfer characteristic of an operational amplifier.

The input signal range over which linear amplification is achieved is so small that for large input signals the amplifier output can be assumed to have only two states corresponding to the saturation levels. For all practical positive inputs the output is at $+V_{sat}$, for all practical negative inputs the output is at $-V_{sat}$.

This feature can be used to compare the levels of two signals applied to the differential inputs and derive an output signal indicating which is the greater. Output levels of $+V_{sat}$ and $-V_{sat}$ may not be convenient and it is common to modify them by adding limiting diodes. For interfacing to digital circuits, levels of about 0 V and +5 V are usual and figure 6.28 shows a passive limiting circuit with a Zener diode defining output levels of about +5 V and −0.6 V which would be suitable.

The speed of a general purpose operational amplifier is too slow for many

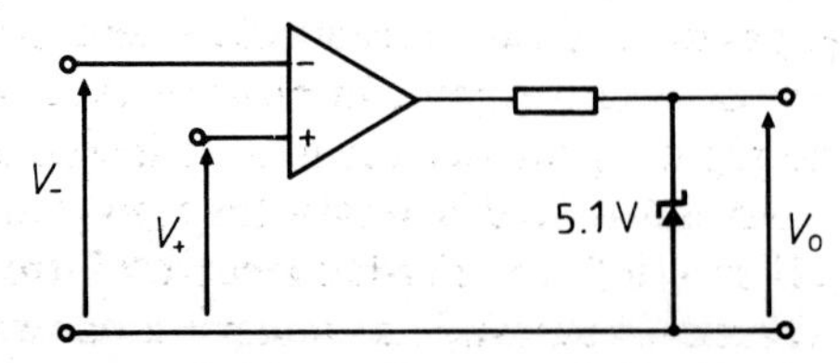

Figure 6.28 Limiting of the output levels of an operational amplifier.

comparator applications and special comparators are available. These are integrated circuit differential amplifiers, but with reduced gain, very fast response and output levels designed to interface directly with digital circuits.

Positive feedback and hysteresis

Comparators without feedback operate very satisfactorily with large input signals. However, small inputs may cause ambiguity if the comparator output falls in the linear range of the transfer characteristic. Another problem can be caused by the noise which is inevitably present in all practical signals. Consider an example where a comparator is used to detect when a rising signal applied to the non-inverting input exceeds zero. In the ideal case with a noise-free signal, as shown in figure 6.29, a clean transition occurs at the output as the input level passes through zero. But if there is noise, as in figure 6.30, the output may alternate several times between the two states.

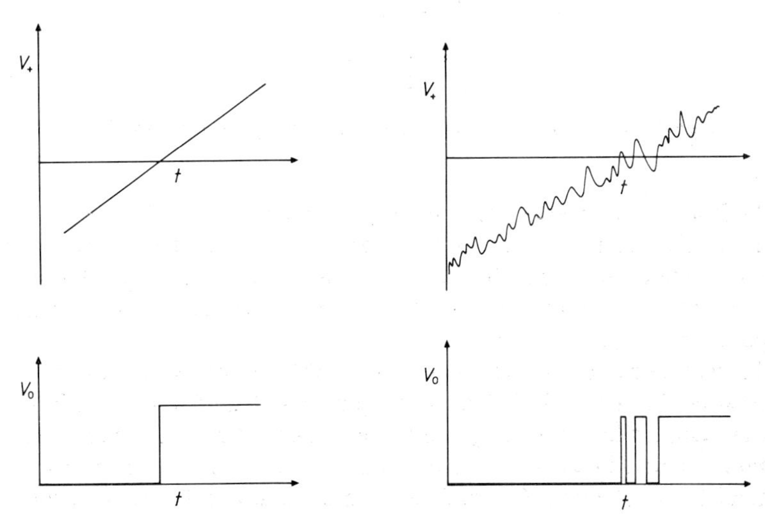

Figure 6.29 Output waveform of a comparator with an ideal input signal.

Figure 6.30 Output waveform of a comparator with a noisy input signal.

Both of these problems can be overcome by the application of positive feedback to the comparator. Negative feedback, where the feedback signal opposes the input signal, reduces gain. In positive feedback the feedback signal reinforces the input signal and the gain is increased. Only a small amount of positive feedback is needed to raise the high gain of a comparator or operational amplifier to infinity. Further feedback introduces hysteresis, where the transfer characteristic exhibits memory as shown in figure 6.31.

There are two configurations possible with positive feedback over a

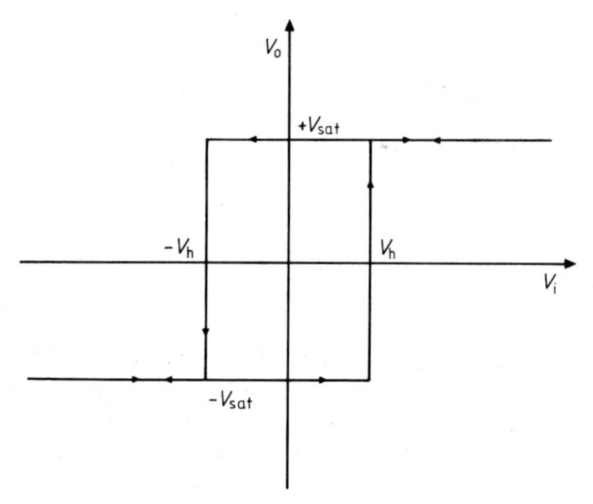

Figure 6.31 Hysteresis introduced in a differential amplifier by positive feedback.

differential amplifier just as there are with negative feedback. Figure 6.32 shows the non-inverting version. The width of the hysteresis loop can be derived by considering the point at which the differential input to the comparator changes sign. If the output is at $+V_{sat}$ in figure 6.32, the voltage at the input V_h must be $-(R_1/R_2)V_{sat}$ to make the differential input voltage zero.

When the output undergoes a transition, say from $-V_{sat}$ to $+V_{sat}$ at an input of V_h, the input must fall to $-V_h$ before the reverse transition can occur. So noise of peak to peak amplitude less than $2V_h$ can be tolerated without giving rise to false output transitions. Figure 6.33 illustrates the improvement. Naturally the hysteresis presents a limit to the accuracy of the comparison but this is a small price to pay for the effective removal of the noise on the signal.

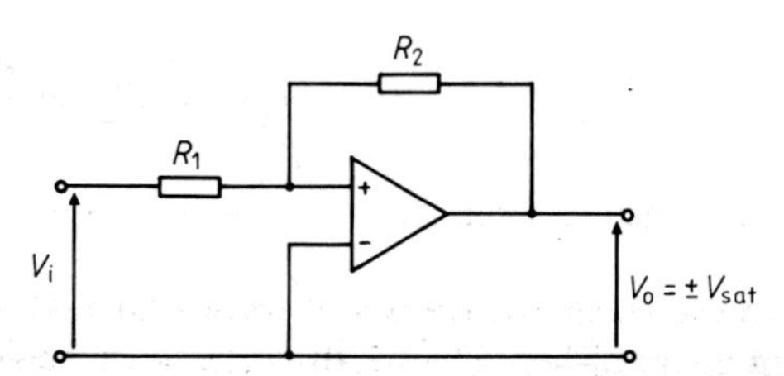

Figure 6.32 Non-inverting configuration of a differential amplifier with positive feedback.

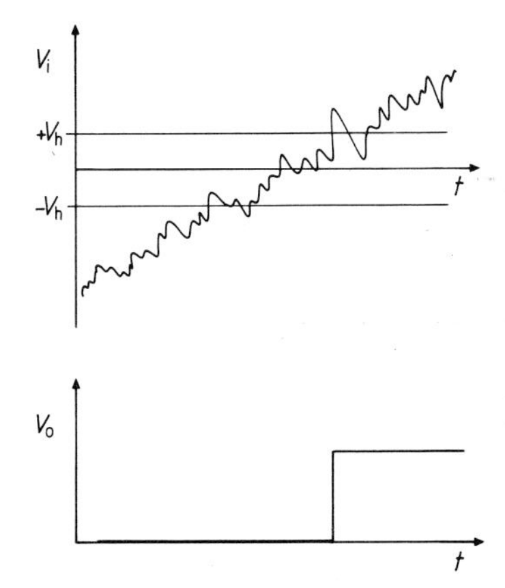

Figure 6.33 Output waveform from a comparator with positive feedback. The noise on the input signal is suppressed if its peak to peak amplitude is less than the hysteresis width.

6.6 Transistor Switches

Interfacing between analogue and digital signals and between digital signals of different levels is a common requirement. Comparators are useful elements, but are unnecessarily precise for some applications and deliver insufficient power for others. Many of these applications can be filled by discrete transistors used as switches.

Figure 6.34 shows a basic switch circuit using a BPT and a resistive load. Like a comparator, this switch can be regarded as an overdriven amplifier. For values of V_i well below the turn-on point of about 0.6 V, there is no base current, the transistor is cut off and hence the collector voltage rises to the power supply voltage. If the input voltage is raised well above 0.6 V, a large base current flows and the transistor is turned hard on. So the collector voltage falls to the saturation level.

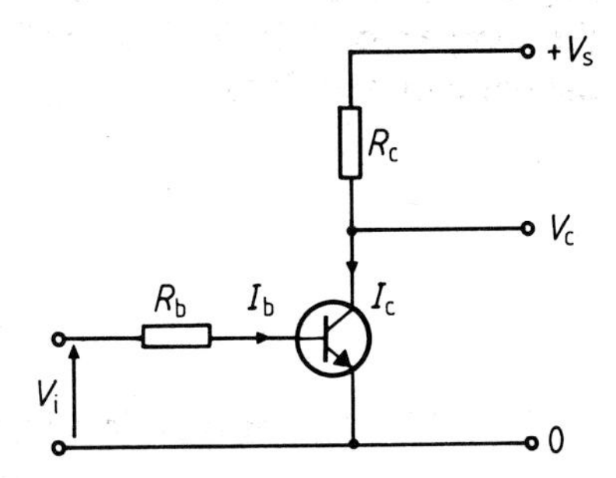

Figure 6.34 Bipolar transistor as a switch.

In between these extremes, the transistor passes through the linear region where it behaves as an amplifier. The width of the linear region is determined by the values of R_b, R_c, the supply voltage V_s and the current gain β of the transistor.

There is little difficulty in establishing the OFF condition. Any input voltage below the turn-on voltage of the transistor will guarantee that I_b and I_c are zero. But saturation is not quite so easy to define. If the current gain of the transistor remained constant, saturation would be reached when $I_c = V_s/R_c$, $I_b = I_c/\beta$ and $V_i = I_b R_b + V_{be}$. But as the transistor nears saturation its current gain falls and it is necessary to provide more base current than these expressions would indicate. It is usual to allow a 'safety factor' of two, and hence to ensure satisfactory saturation and force the transistor to the ON state, the value of I_b is normally taken as $2I_c/\beta$, where the current gain is the value in the active region.

Modern BPT, especially those optimised during manufacture for switching operations, show very good saturation. The saturation voltage may be below 0.1 V and in many cases this can be neglected in comparison with the other circuit voltages. A useful feature of BPT is that the saturation voltage is nearly always below the turn-on point, so that one transistor switch can directly drive another as shown in figure 6.35.

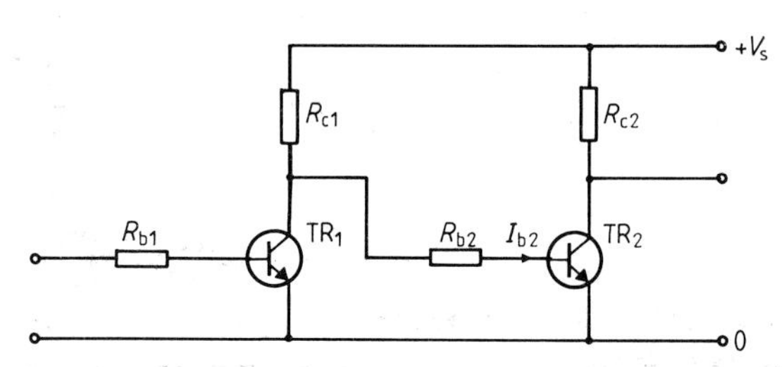

Figure 6.35 Cascaded transistor switches.

If TR_1 is ON then V_{c1} will be low enough to ensure that TR_2 is OFF. In many cases R_{b2} is omitted, and the base current of TR_2 in the ON state is defined by R_{c1} at $I_{b2} \approx V_s/R_{c1}$.

Power dissipation and efficiency

Transistors are very efficient switches. Power dissipation in a BPT is approximately the product of the collector current I_c and the collector–emitter voltage V_{ce}. When the transistor is OFF I_c is negligible, and so is the power dissipation. In the ON state, V_{ce} is small and thus the power dissipation is also small. The power dissipated in the load R_c is V_s^2/R_c if the saturation voltage can be neglected. A low power switching transistor rated at only a few hundred milliwatts can easily control load powers of tens of watts. Moreover, power transistors are available with rated dissipations of hundreds of watts that can control powers of many kilowatts.

Inductive Loads

Transistors are often used to switch inductive loads such as relays. When the current through an inductance changes a back EMF is generated. A small transistor can switch a current of a 100 mA or so in a time much less than a microsecond. If this rate of change of current is forced through an inductance of 100 mH, a typical value for a relay, the back emf exceeds 10 kV. This is far above the breakdown voltage of all small transistors and would be fatally destructive.

It is usual to suppress this voltage spike with a diode limiter circuit as shown in figure 6.36. The diode protects the collector and is returned to the most convenient reference voltage, the power supply. The negative going spike generated when the transistor switches ON is controlled by the reduction in current gain as the transistor approaches saturation, which slows down the rate of current rise. When the collector current is switched OFF the positive going induced spike drives the diode into conduction and the collector voltage cannot rise above ($V_s + V_f$).

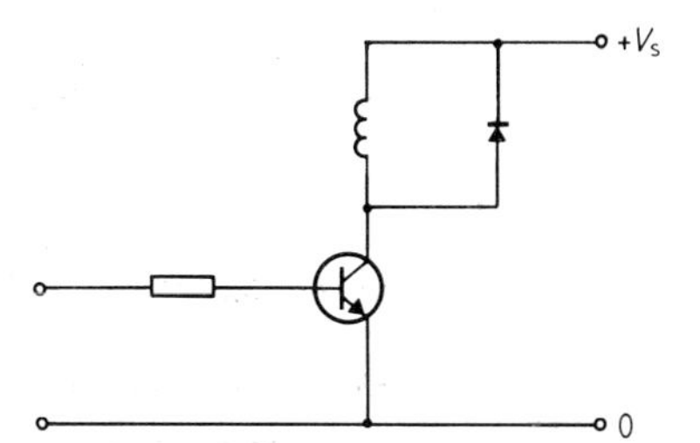

Figure 6.36 Diode limiting for a transistor switching an inductive load.

MOST and analogue switching

MOST can be used as switches in just the same way as BPT, except that operation is by direct control of the gate to source voltage. The very high input resistance of MOST is a useful property, but the effective drain–source resistance in the ON state is usually higher than for a BPT.

A particular property of MOST is that they can be used to switch analogue signals. The conduction process in a MOST operates for either polarity of drain–source voltage, as shown in a plot of the characteristics near the origin for an enhancement n channel device (figure 6.37). The gate–source voltage effectively modulates the resistance of the channel; this can change from many megaohms in the OFF state to a few ohms in the ON state. The channel resistance—the slope of the I_d against V_{ds} curve—is not a function of V_{ds} for low values of V_{ds}.

So a MOST can be used as a voltage controlled resistance provided that the drain–source voltage is kept small. Figure 6.38 shows a simple application

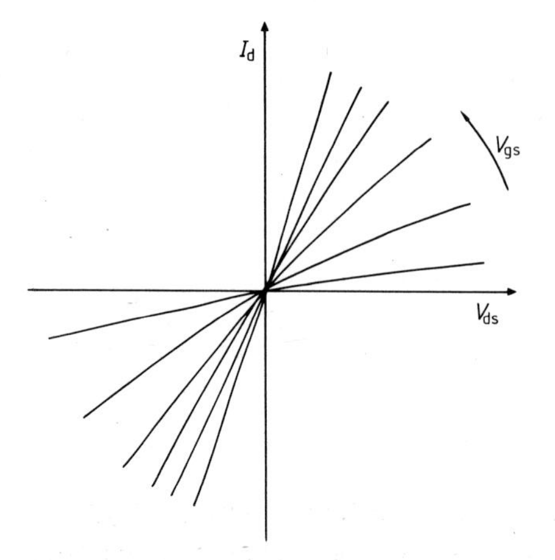

Figure 6.37 Voltage–current characteristics of an n channel MOST near the origin.

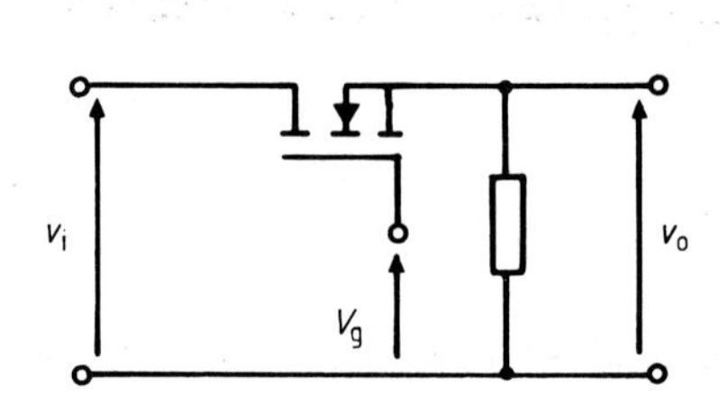

Figure 6.38 MOST as an analogue switch.

where an n channel MOST forms the upper section of a potential divider. When the gate voltage is low, the transistor presents a very high resistance and the output signal v_o is nearly zero. When the gate voltage is high, the device has low resistance and almost all of the input signal v_i is transmitted to the output.

Figure 6.39 shows a modification of the operational amplifier summer where the input signals can be selected or deselected by controlling the gate voltage of the appropriate n channel MOST switch in the input circuit.

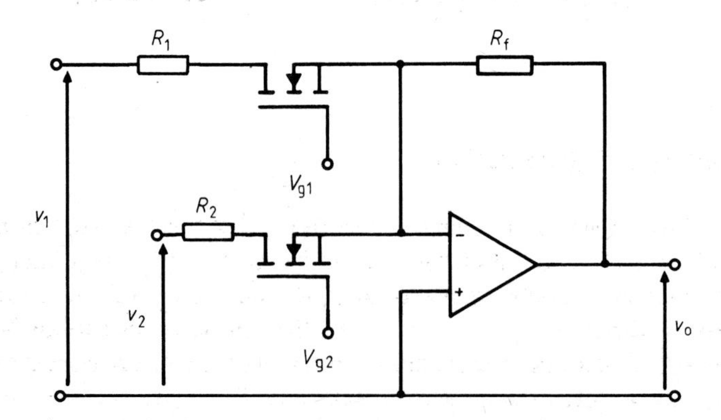

Figure 6.39 Active summation with input selection.

7

Digital Electronics

Digital electronics processes signals restricted to a finite number of values. Some signals are inherently digital. For example, the output from the two-way switch in figure 7.1 can only be at 0 V or 5 V. Naturally, signals from other digital systems are themselves digital. However, most signals are continuous and analogue, and before they can be processed by digital circuits they must be translated into digital form by making them discrete both in time and amplitude. They can then be encoded as a set of digits.

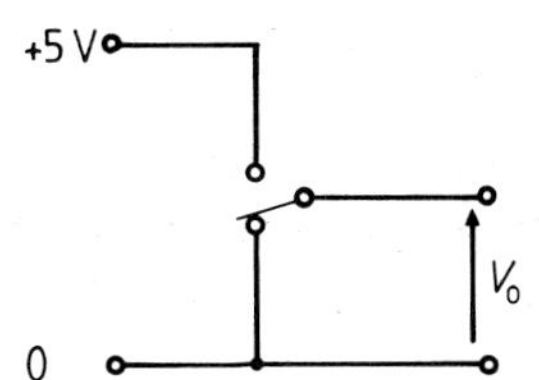

Figure 7.1 Two-way switch.

7.1 Sampling and Quantisation

The sampling operation is illustrated in figure 7.2 for the usual case where the samples are taken at regular intervals. The result of the sampling process is a sequence of signals corresponding to the values of the continuous waveform at times t_0, t_1, t_2 . . . . In practice the samples cannot be taken instantaneously, and the values are an average over a short period whilst in between the sample times the signal is undefined. It will be clear that sampling modifies the information present in the original signal. However, it is intuitively obvious that straight line interpolation between the samples gives a close approximation to the original continuous signal, provided that the time between samples is short compared with the time scale of detail in the signal. This is equivalent to saying that the sampling rate must be high

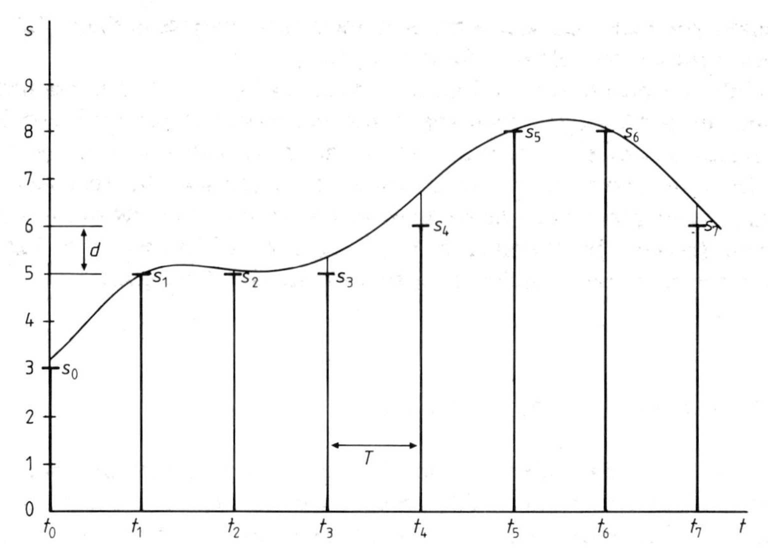

Figure 7.2 Sampling and quantisation of an analogue signal.

compared with any of the frequency components in the signal. More formally, this idea is expressed in the sampling theorem.

The sampling theorem

This theorem states that when a signal is sampled at a rate f_s, (a sampling interval of $1/f_s$), all frequency components in the signal up to $f_s/2$ are preserved. Frequency components above $f_s/2$ become indistinguishable from their reflections about $f_s/2$. (A frequency component $f_s/2+f$ produces the same contribution to the output as one at $f_s/2-f$.) In most cases this aliasing effect is undesirable and the input signal must be low-pass filtered to remove components above $f_s/2$. Sometimes the sampling rate will be chosen to be sufficiently fast that the aliases are negligible and then no filter is needed.

We now have a signal which is discrete in time, but is still continuous in amplitude. Quantisation is the process of reducing the infinite range of possible values to a finite number of allowed values. Normally the quantisation interval d is constant, but a variable interval is sometimes used to obtain better resolution at small signal values. Two forms of quantisation are possible: either the continuous value can be rounded to the nearest allowed discrete value, or it can be truncated to the value immediately below. Truncation gives a quantisation error between 0 and d, but the maximum error with rounding is $d/2$. Despite the advantage of rounding, truncation is usually adopted because it is much simpler. The samples in figure 7.2 have been truncated to yield the sequence s_0, s_1, s_2 . . . .

Only ten quantisation values are shown in figure 7.2. The number of values needed for a particular application depends on the accuracy required. A representation to 0.1% is enough for most instrumentation purposes, in which case about 1000 values would be necessary. Alternatively, the quantisation process can be regarded as introducing a nearly random error signal (noise). The interval must be chosen sufficiently fine that the quantisation noise is small compared with detail in the signal.

7.2 Logic States and Binary Notation

Circuits can be made which operate directly with ten signal values. They have some advantage because they are convenient for processing decimal numbers. But binary circuits, using just two voltage states to represent the signal, have become almost universal. The advantage of binary circuitry is that it is much easier to distinguish between two states than to separate a larger number. In many binary circuits the two states correspond to OFF and ON in an electronic switch, that is simply the presence or absence of a signal voltage or current. Considerable variations in component values or circuit operating conditions are possible without any effect on the ability of the circuits to separate the two values. Noise and spurious signals, whether generated within the circuits or induced, are also unimportant provided that they never become so large that the two states cannot be reliably distinguished.

Many of the signal processing operations provided by simple binary circuits correspond to logic operations on TRUE and FALSE values, and these circuits are often referred to by their logic functions even when being used for other purposes. Also, the binary number system uses only the digits 0 and 1, so these symbols are often used to describe the two possible states of a digital signal. It is common to assign logic 0 to the state represented by a low voltage at or near 0 V and logic 1 to some higher voltage.

Signals which are inherently two-state, such as those from the switch in figure 7.1, can be directly processed by binary circuits, but some form of coding is necessary before a signal which has been quantised to a larger number of values can be processed. One obvious arrangement is that of figure 7.3 for a ten-value system. Each value is indicated by activating just one of the ten binary output signals. One possible allocation of the output lines and the corresponding output codes are shown. However, this simple arrangement requires as many output lines as values of signal and this becomes unwieldy for the hundreds or thousands of values required in many signal processing systems. More efficient coding is possible.

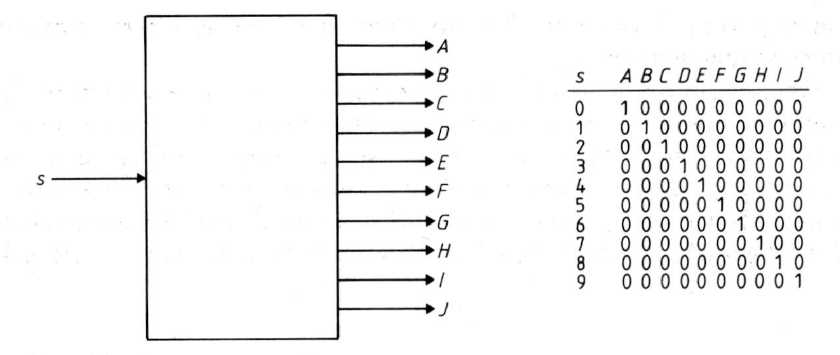

s	A	B	C	D	E	F	G	H	I	J
0	1	0	0	0	0	0	0	0	0	0
1	0	1	0	0	0	0	0	0	0	0
2	0	0	1	0	0	0	0	0	0	0
3	0	0	0	1	0	0	0	0	0	0
4	0	0	0	0	1	0	0	0	0	0
5	0	0	0	0	0	1	0	0	0	0
6	0	0	0	0	0	0	1	0	0	0
7	0	0	0	0	0	0	0	1	0	0
8	0	0	0	0	0	0	0	0	1	0
9	0	0	0	0	0	0	0	0	0	1

Figure 7.3 Encoding of a ten-level signal.

7.3 Binary and BCD Encoding

A pure binary encoding scheme, where each of the n output lines can take either the 0 or 1 states, allows 2^n distinguishable combinations corresponding to 2^n discrete values. The ten-value example now needs only four output lines as shown in figure 7.4. For higher numbers of values the saving is more dramatic, ten output lines allowing $2^{10} = 1024$ combinations. An orderly assignment of the binary codes, where the binary number formed by the code corresponds to the signal value, is convenient and common but it is not essential.

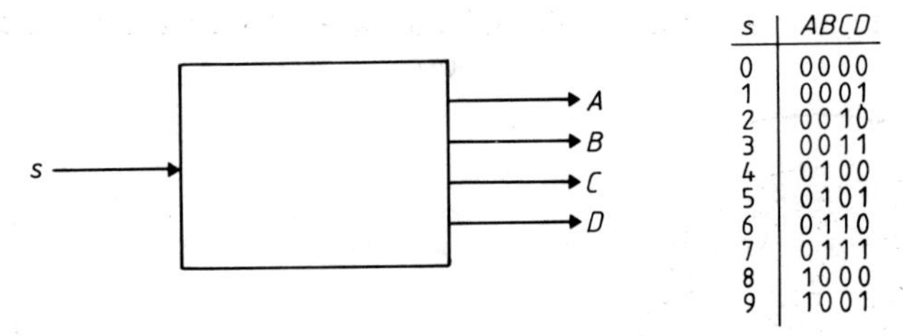

s	ABCD
0	0000
1	0001
2	0010
3	0011
4	0100
5	0101
6	0110
7	0111
8	1000
9	1001

Figure 7.4 Pure binary encoding of a ten-level signal.

Although pure binary encoding is convenient for digital circuits, it is cumbersome for humans. We prefer to count on our fingers and so use the decimal number system. Binary coded decimal (BCD) is a compromise between pure binary and decimal that is widely used in instrumentation. Each digit in the decimal number is encoded in pure binary and needs four binary digits (bits) as illustrated in figure 7.5. The binary codes corresponding to the decimal numbers 10–15 are not used, so BCD must be less efficient than pure binary.

Where pure binary is used, octal or hexadecimal representation of the binary numbers is often convenient. In octal (base eight), the bits of the binary word are grouped in threes from the least significant end and described by a digit in the range 0–7. In hexadecimal (base 16), the grouping is in fours and the symbols 0–9 and A–F are used to represent the values 0–15. Figure 7.6 shows the relation between binary, octal and hexadecimal.

Decimal	1	5	9
BCD	0001	0101	1001

Figure 7.5 BCD representation.

	Code
Decimal	159
Binary	10011111
Octal	10,011,111 = 237_8
Hexadecimal	1001,1111 = $9F_{16}$

Figure 7.6 Relations between decimal, binary, octal, and hexadecimal number representations.

Gray codes

There is a further type of code commonly encountered in instrumentation. Angular or linear displacement can be converted into digital form through a set of switches which can be mechanically or optically actuated. Figure 7.7 shows an angular encoding disc with a pattern of contacts arranged to produce three-bit pure binary output code. However, this angular transducer has a defect. When all three digits change at the transition between codes 011 and 100, inevitable misalignment of the contacts means that the three digits cannot change simultaneously, and invalid codes will be produced. Figure 7.8 shows one of the several possible transient code

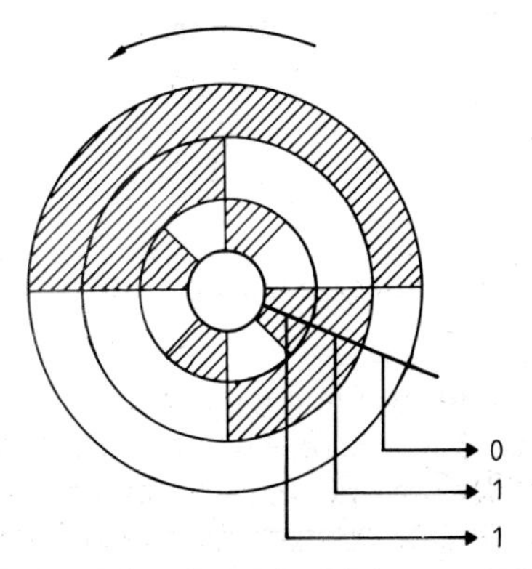

Figure 7.7 Three-bit binary angular encoder.

Code	
0 1 1	Valid code
0 0 1	Invalid
0 0 0	codes
1 0 0	Valid code

Figure 7.8 Invalid codes at the transition from 011 to 100.

sequences that might occur during the transition. Similar effects occur at any code boundary where more than one bit changes.

In some cases it is possible to arrange that the output code is sensed only at positions away from the contact boundaries. Or the invalid codes may be produced for a sufficiently short time that they can be ignored. Yet an alternative coding scheme, where only one bit of the code can change at any time would remove the problem completely. These codes are called Gray codes and an example suitable for this particular problem is shown in figure 7.9.

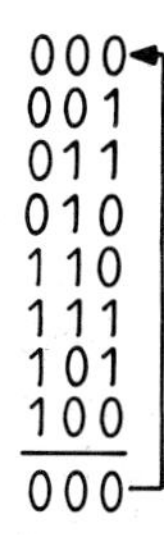

Figure 7.9 Eight-state Gray code.

Text codes

The code representations above have all assumed that the information is numeric. But binary words can represent other information too. Text characters are inherently digital signals that can be encoded in several ways. The American standard code for information interchange (ASCII) uses a seven-bit word to represent alphanumeric characters and control codes for printers and displays. Text information in digital form can be manipulated and processed by digital circuits in the same way as numeric information.

Parallel and serial operation

In all these examples digital information has been considered in the form of words—sets of bits—in which all of the information is present simultaneously. It is available on a group of lines, often called a bus, in parallel. An alternative form of presentation of a digital word is to offer the bits one at a time in sequence on a single line, known as serial operation. Parallel operation achieves greater signal processing speed at the expense of more circuitry and more interconnection whilst serial operation is appropriate where high speeds are not essential and the extra costs of parallel working cannot be justified.

7.4 Logic Gates

The simplest operation that can be performed on a digital signal is inversion or negation. An inverter changes a logic 1 into logic 0 and vice versa and provides the logic NOT function. Figure 7.10 shows an inverter operating on a signal A. The small circle at the output of the triangular inverter symbol indicates inversion.

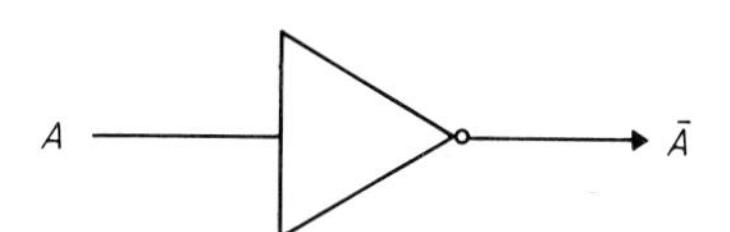

Figure 7.10 An inverter.

The next simplest operation combines two digital signals to produce a third as shown in figure 7.11. The input signals A and B can each take either of the values 0 and 1. So there are four possible combinations of the input variables. For each of these input combinations the output can also take either the value 0 or 1. Hence in principle there are 16 (i.e. 2^4) possible operations on the input signals that such a circuit might perform. These are shown in tabular form in figure 7.11.

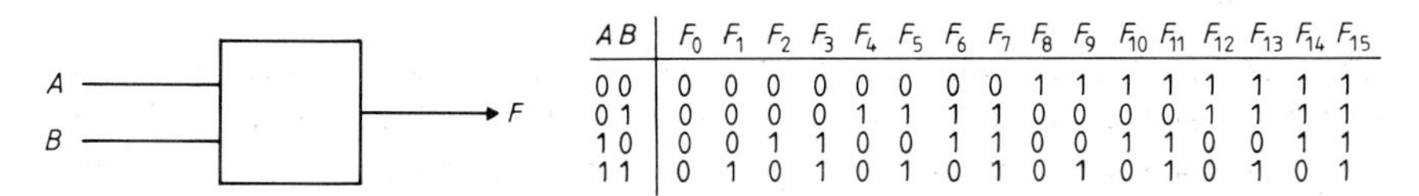

AB	F_0	F_1	F_2	F_3	F_4	F_5	F_6	F_7	F_8	F_9	F_{10}	F_{11}	F_{12}	F_{13}	F_{14}	F_{15}
00	0	0	0	0	0	0	0	0	1	1	1	1	1	1	1	1
01	0	0	0	0	1	1	1	1	0	0	0	0	1	1	1	1
10	0	0	1	1	0	0	1	1	0	0	1	1	0	0	1	1
11	0	1	0	1	0	1	0	1	0	1	0	1	0	1	0	1

Figure 7.11 Two-input logic operations.

Some of these operations are trivial. For example, F_0 gives a constant output independent of the input variables, and F_3 just repeats the value of A. However, there are some interesting operations which are sufficiently general that they can be extended to the case of more than two input variables. The operation F_1 can be summarised in words. The output becomes 1 when (and only when) A is 1 AND B is 1. Thus it is the logic AND function. F_7 corresponds to the output becoming 1 when A is 1 OR B is 1 OR both are 1 and is the logic OR function. F_6 is also sometimes considered to be a fundamental logic function. It is called the EXCLUSIVE OR because the output is 1 when either A is 1 or B is 1, but excludes the case when both A and B are 1. The previous function F_7 can be called the INCLUSIVE OR to distinguish it from F_6 if there is any possibility of confusion.

F_{14} and F_8 are inversions of F_1 and F_7 and correspond to NOT AND and NOT OR, usually abbreviated to NAND and NOR. AND, OR, their inverted forms NAND

and NOR, and the NOT function can be regarded as the basic building blocks from which digital systems are made and the symbols for these are shown in figure 7.12.

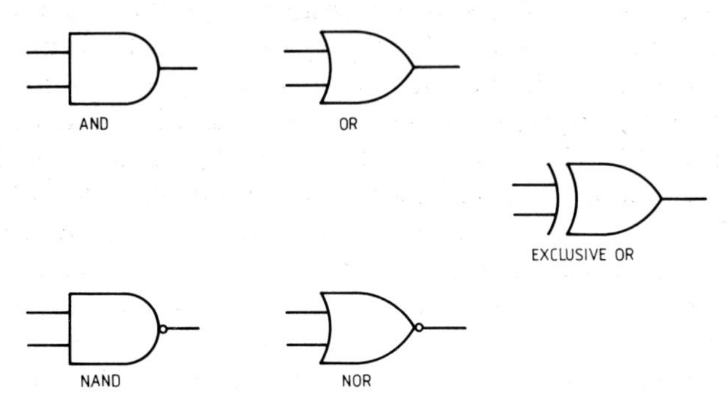

Figure 7.12 Two-input gates.

One of the simplest uses of the AND circuit is to control the flow of a digital signal. Suppose a signal—which will consist of a stream of 0s and 1s—is presented to input A. Now if input B is held at logic 1, inspection of figure 7.11 shows that the output just follows the value of A. But if B is held at logic 0 the output is forced to 0 also. So the signal stream at A is either transmitted or blocked according to the value of the control signal on B, as shown in figure 7.13(a). Examination of the possible states of F_7 shows that an OR circuit can be regarded as combining two signal streams as shown in figure 7.13(b).

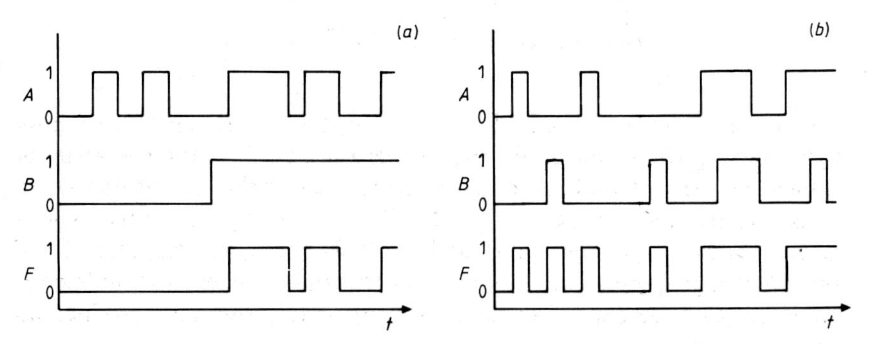

Figure 7.13(a) AND, (b) OR operations on digital signal streams.

These operations on signal streams lead to the use of the term 'gates' to describe the basic logic circuits, even when they are being used for other purposes.

7.5 Practical Gates—TTL

There are many electronic circuits that can be designed to perform logic operations on digital signals, but one circuit family has been predominant for more than a decade. Transistor–Transistor Logic (TTL) was the first logic family successfully to exploit the properties of silicon integrated circuit technology and is still in use today in its original form plus several new variants. TTL gates use bipolar transistors as switches in a manner similar to that outlined in §6.6. However, the inputs to a TTL gate do not go to separate transistor switches but to a single transistor with multiple emitters.

Figure 7.14 shows the general form of a three-input TTL gate. The unusual input configuration can cause problems for the unwary. If any of the inputs are held at or near ground potential, the current flowing down R_1 passes through the forward biased base–emitter junction and out of that input terminal, turning TR_1 ON and TR_2 OFF and thus giving a high gate output. Only when all the inputs are held high will current flow down R_1 through the now forward biased base–collector junction of TR_1 into the base of TR_2, switching in ON to make the gate output low.

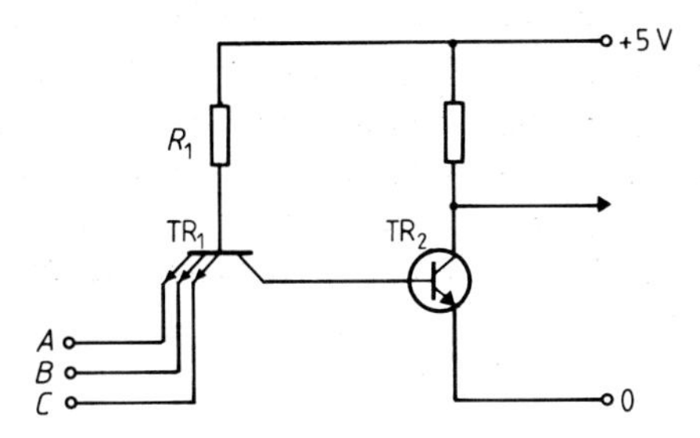

Figure 7.14 Outline three-input TTL NAND gate.

Practical implementations of this circuit include buffering of the output signal and arrangements to raise the voltage at which TR_2 turns ON. An input voltage between 0 V and about 0.8 V will ensure that the current flow down R_1 is diverted away from the base of TR_2. So any signal in this range is recognised by the gate as signifying logic 0. Voltages between about 2.4 V and 5 V will guarantee that TR_2 is ON and this range corresponds to logic 1. Input voltages between 0.8 and 2.4 V are ambiguous and produce an undefined output.

It is a natural, but quite unjustified, expectation of many users that shorting an input terminal to ground will result in no current flow. These TTL gate inputs supply about 1.6 mA under these conditions. Also, if a gate input is left unconnected (floating), it corresponds to logic 1 not logic 0 as might be supposed. Perhaps it should be noted here that gate inputs left floating are liable to pick up spurious signals (noise) which may be large enough to cause errors. (It is very bad practice to make use of this floating feature of TTL gate behaviour in defining a logic 1 at an input.)

The basic TTL gate performs the NAND function. Other functions such as NOR, AND and OR can be produced by extensions of the circuitry. Alternatively, these functions can be produced by using combinations of NAND gates as we shall see later.

Many TTL gates can be fabricated on a single chip and encapsulated in the same package. The limitation is the number of external pin connections required. For example, the first member of the TTL gate family, the 7400, has four two-input NAND gates in a 14-pin package. Three three-input gates (7410), two four-input gates (7420) or a single eight-input gate (7440) can be accommodated in the same size package. A wide variety of packages and functions, some much more complex than these simple gates, is now available.

Standard TTL has now been supplemented by variants which offer different combinations of speed and power consumption. At present the most widely used of these is the low power Schottky (LSTTL) range which provides the same speed as the original family but consumes only about a tenth of the power.

CMOS

Gates based on complementary p channel and n channel MOST switches have become popular within the last few years for two reasons. The power consumption is very low, a considerable advantage for battery powered equipment, and the circuits are more tolerant of variations in power supply voltage, whereas the TTL families require a well-defined supply for satisfactory operation. The basic gates available in CMOS are similar to those obtainable in TTL form with AND, NAND, OR and NOR multigate chips available. Figure 7.15 shows a basic CMOS NOR gate.

When the inputs *A* and *B* are both low, the p channel devices are ON and the n channel devices are OFF. So the output must be high. If either *A* or *B* is taken high, one of the series p channel transistors turns OFF and one of the parallel n channel transistors turns ON. The output then goes low. In either output state, there is at least one non-conducting MOST in the path between the power trials, so only a very small leakage current flows. Only during the switching transition does the gate draw any substantial supply current.

Unfortunately the voltage values corresponding to logic 0 and 1 are not the same as with TTL. In standard CMOS, signal levels between 0 V and about a third of the power supply voltage are taken to be logic 0. Logic 1 is represented by signal levels above about two thirds of the power supply voltage. The absolute value of the power supply can vary over a wide range with 3 V to 15 V being normal. It is possible to arrange that signals from TTL

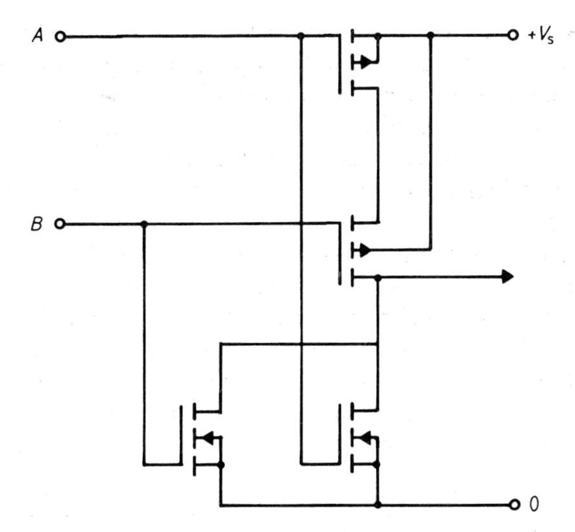

Figure 7.15 Outline two-input CMOS NOR gate.

gates can drive CMOS gates and vice versa, but some care is needed. In particular the CMOS supply voltage must be 5 V to correspond with the TTL supply, and some level shifting is necessary when driving CMOS from TTL because a TTL logic 1 output may fall as low as 2.4 V. A modified CMOS gate family has been introduced which offers direct compatibility with LSTTL levels and has similar speed.

Other logic families

Other types of circuitry can be used to form logic functions. Emitter coupled logic (ECL) uses bipolar transistors and is very fast and its use is largely confined to very high speed computers. Gates made with only one polarity of MOST (NMOS or PMOS) may be used in the LSI circuits discussed in Chapter 8 but their properties make them unsuitable for use as individual gates.

Speed and power

Logic gates react very rapidly to input signals. After a change in the input variables there is a delay of about 10 ns before a TTL gate assumes the output state corresponding to the new input combination. Standard CMOS gates tend to be a little slower. It may be difficult to appreciate how short these times are compared with the time scale of events in the non-electronic world. For comparison, a supersonic airliner travels only a few micrometres in 10 ns. Light manages 3 m.

For many purposes these delays are negligible, even when signals pass through several successive gates. However, they become important when processing high speed signals. Such signals are commonly encountered in computing systems or they might come, for example, from a digitised television picture.

An LSTTL gate dissipates between 2 and 3 mW, that is it draws approximately 500 μA from a 5 V supply. CMOS gates draw almost no current at all except when switching from one logic state to the other. Typically a CMOS NOR gate requires a few nanoamps when quiescent rising to about 1 mA when switched at a rate of 1 MHz.

Interconnection and loading effects

In §4.6 we saw that signal transfer between linear amplifier stages can be described in terms of input and output resistance. Gates are non-linear elements and the input and output resistances depend on the logic state. Also, as we have seen for the TTL gate, inputs may be active. Hence it is better to think in terms of loading rules determining how gates may be interconnected.

These rules are quoted by the device manufacturer and allow for extreme variations in supply voltage, device tolerances, ambient temperatures etc. A standard TTL gate output is guaranteed to drive up to 10 similar gate inputs whilst special high power buffers are available which can drive up to 40 inputs.

Although binary circuits have only two valid logic states, some gates are provided with an enable/disable input which controls the connection between the internal logic circuits and the output buffers. When disabled the outputs are effectively disconnected and become free to float. This is regarded as a third state and gates with three-state (tristate) buffers can have their outputs connected to a common line. The gate whose tristate output is enabled defines the logic level carried on the line. Naturally, to avoid contention, not more than one gate connected to a common line must be enabled at any time.

7.6 Truth Tables, Boolean Algebra and Karnaugh Maps

A truth table is a concise way of summarising the relations between the input and output signals in a digital system. Figure 7.16 is a truth table showing the properties of all of the basic two-input gates we have defined. The major uses of truth tables are for describing and analysing the behaviour of combinational logic, that is, digital systems where the output depends only on the current state of the input signals.

AB	AND	OR	NAND	NOR	EXOR
00	0	0	1	1	0
01	0	1	1	0	1
10	0	1	1	0	1
11	1	1	0	0	0

Figure 7.16 Truth tables for the basic two-input gates.

Figure 7.17 shows a general combinational logic block with m inputs and n outputs. The behaviour of this system can be completely described (neglecting the delays mentioned above) by a truth table which shows the state of each of the n outputs for every combination of the m inputs. So there will be n (2^m) entries in the table. For systems with large numbers of inputs the table becomes very large and it may be possible to abbreviate the table where repeated or regular entries occur. In such cases it must be remembered that all input combinations have to be accounted for, either explicitly or implicitly.

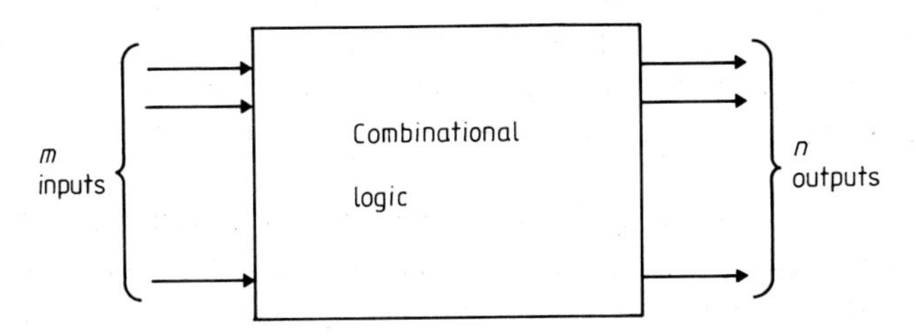

Figure 7.17 General combinational logic system with m inputs and n outputs.

Occasionally the truth table of a combinational logic system that we require will happen to coincide with that of one of the standard gates. It is then obvious how to make the system. However, usually this is not so and we require a general method of designing a network of basic gates which will perform the desired function. Consider first how we might produce the function F_2 in figure 7.11. Although this is not the AND function, it is very similar. The output is 1 when A is 1 AND when B is NOT 1. So an AND gate which operates on A and on the inverse of B will give F_2 as shown in figure 7.18.

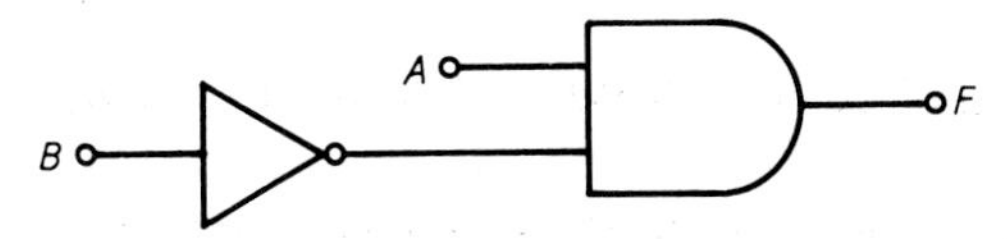

Figure 7.18 The circuit needed to produce the function F_2 (figure 7.11).

As another example, let us consider the design of some logic to produce a two-way switching effect. There are two inputs, A and B, and an output F. F is to be OFF (0) when A and B are OFF, and to become ON (1) if A is switched

ON or if B is switched ON. However, if A and B are both ON then F must go OFF (this is F_6, the EXCLUSIVE OR function, but we shall assume that an EXCLUSIVE OR gate is not available). The behaviour is described in figure 7.19, where it is shown that the output F_6 can be obtained by an OR operation on two simpler functions P and Q. We recognise P and Q as being AND functions with appropriate inversions of their input signals, and so the network of figure 7.20 must realise the desired function.

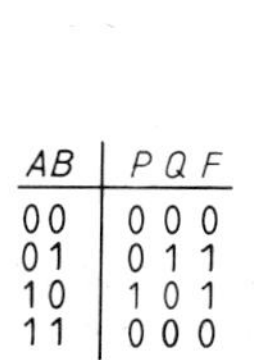

AB	PQF
00	000
01	011
10	101
11	000

Figure 7.19 Truth tables for two-way switching.

Figure 7.20 Logic realisation of two-way switching.

Boolean algebra

It is possible to express the two-way switch problem in words and to deduce the network directly by seeing that F is 1 when A is NOT 1 AND B is 1 OR when A is 1 AND B is NOT 1.

An equivalent but more concise way of expressing the relation between A, B and F is to use an algebra which can be applied to logic problems. Boolean algebra operates on switching circuits and is widely used for the definition and analysis of combinational logic.

In Boolean algebra, symbols represent the logical operations of AND, OR and NOT. Unfortunately, there are different symbol sets in use by mathematicians and electronic engineers. In engineering notation AND and OR are represented by the $\times$ and $+$ symbols as used for multiplication and addition in ordinary algebra. The $\times$ symbol is often omitted and understood to be implicitly present in a term such as

$$\mathrm{ABC} = \mathrm{A} \times \mathrm{B} \times \mathrm{C}.$$

The NOT operation is indicated by a bar over the variable.

Our problem can then be expressed as

$$F = \bar{A}B + A\bar{B} \tag{7.1}$$

Boolean algebra obeys many of the rules of ordinary algebra, but with some special rules of its own. Some of the most useful relations are summarised in table 7.1.

Table 7.1 Some useful Boolean relations.

$AB = BA$	$A0 = 0$	
$A + B = B + A$	$A1 = A$	
$A(B + C) = AB + BC$	$AA = A$	
$A + 0 = A$	$A\bar{A} = 0$	
$A + 1 = 1$	$\overline{AB} = \bar{A} + \bar{B}$	De Morgan's theorem
$A + A = A$	$\overline{A + B} = \bar{A}\bar{B}$	
$A + \bar{A} = 1$	$A + \bar{A}B = A + B$	
$A + AB = A$		

Simplification of logic

The AND–OR approach to the solution of the problem above is quite general. All problems in combinational logic can be described by truth tables or by a Boolean sum of products expression like equation (7.1). Both forms lead directly to a network of AND gates whose outputs are combined with an OR gate. However, in many cases this will not be the simplest solution.

This can be illustrated by considering another example, this time with three inputs, A, B and C. The single output F is required to indicate the majority signal of the three inputs. Figure 7.21 shows the truth table and the equivalent Boolean sum of products expression. This problem can be directly realised with four two-input gates, a four-input OR gate and three inverters. But if the output is true for both of the input conditions $AB\bar{C}$ and ABC, the value of C is irrelevant. So these two terms can be reduced to a single term AB. The same argument can be applied to the terms $A\bar{B}C$ and ABC to yield AC and to $\bar{A}BC$ and ABC to yield BC. So the problem can also be realised with just three two-input AND gates and a single three-input OR gate forming $F = AB + AC + BC$, which agrees with an intuitive solution to the problem.

ABC	F
000	0
001	0
010	0
011	1
100	0
101	1
110	1
111	1

$F = \bar{A}BC + A\bar{B}C + AB\bar{C} + ABC$

Figure 7.21 Truth table for the majority decision of three input signals A, B and C.

Formally, Boolean algebra leads to the same answer. We can write

$$AB\bar{C} + ABC = AB(\bar{C} + C) = AB(1) = AB \quad (7.2)$$

where we have used some of the relations from table 7.1. Similarly we can obtain BC and AC. Note that ABC has then been used three times. This is allowed because in Boolean algebra

$$ABC = ABC + ABC + ABC. \tag{7.3}$$

Simplification by direct inspection of the truth table is only possible for rather elementary problems. In most of these cases the simplification is also obvious from consideration of the original problem, at least with hindsight. Algebraic simplification is unwieldy, especially for problems where the output is not defined for all of the input combinations.

CAN'T-HAPPEN *and* DON'T-CARE *conditions*

Many, probably most, practical logic problems have combinations of input variables which are physically impossible (CAN'T-HAPPEN) or for which the value of the output is immaterial (DON'T-CARE). It is usual to indicate the outputs corresponding to these conditions with X in the truth table. Values must be assigned to the X to solve the problem and thus produce a real network of gates. The choice is arbitrary, but it is sensible to select values which lead to the greatest possible simplification. Sometimes the choice is obvious. If the output corresponding to the input combination ABC in figure 7.21 had been an X, it would have been better to choose $X = 1$ rather than $X = 0$.

Karnaugh maps

In more difficult cases, mapping methods provide a rapid and reliable way of simplification for problems with up to about six variables. Above that, computer assistance is almost essential. A Karnaugh map is a two-dimensional representation of a truth table in which each cell corresponds to a row in the table. The maps are arranged so that adjacent cells differ by only a single variable. Figure 7.22 shows possible arrangements for two, three and four variables. Other layouts are also in common use.

In order to satisfy the variable difference requirement, it is necessary to regard the left- and right-hand edges of the three-variable maps as being coincident. This must be extended for the four-variable map to include coincidence of the top and bottom edges also. Larger problems can be tackled with superimposed four-variable maps, two for five variables and four for six variables. The output values for all of the input combinations are entered in the appropriate cells. In practice it is usual to fill in only those cells which contain 1 and X. Empty cells are understood to contain 0.

Now if two adjacent cells contain 1s, this indicates two terms differing in just one variable. These terms can be grouped into a single term with the elimination of that variable using equation (7.2). The procedure can also be applied to adjacent pairs of cells, where four terms can be combined with the

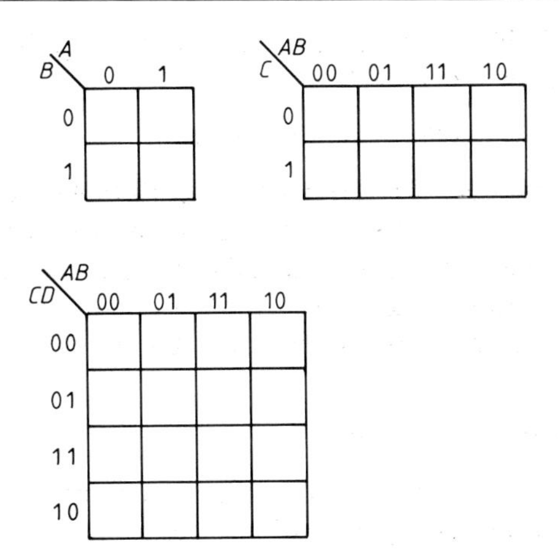

Figure 7.22 Karnaugh maps for 2, 3 and 4 variables.

elimination of two variables. This grouping can be extended indefinitely to rectangular blocks whose size is an integral power of two up to, and including, the entire map. Figure 7.23 shows a mapping of the majority decision problem. The three possible pairings are immediately obvious and are indicated by looping. The solution to the problem can then be written down by inspection as

$$F = AB + AC + BC. \tag{7.4}$$

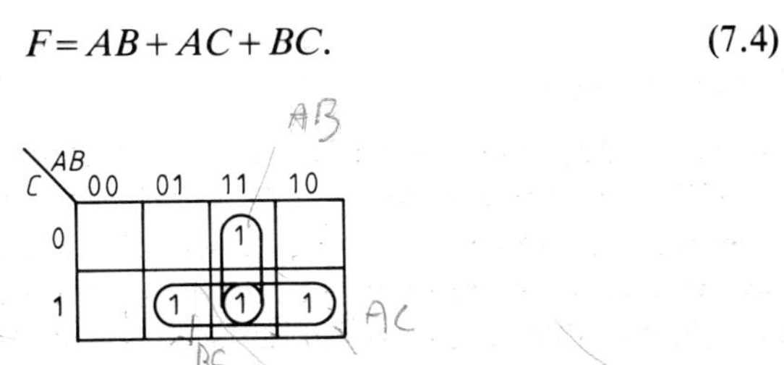

Figure 7.23 Mapping of the majority decision problem. The blank cells are understood to contain 0s.

An example which contains CAN'T-HAPPEN conditions will demonstrate the power of the mapping technique. Suppose we have a motor which positions a platform in a vertical shaft. Four sensors generate signals A, B, C and D, which indicate if the platform is above the particular sensor. Three signals control the motor. M switches the motor on and off, H selects high or low speed and U determines the direction, up or down. The objective is to drive the platform into the region between B and C. If it lies between A and B or between C and D, low speed is used to return it to the centre. Beyond these regions, high speed should be selected.

There are four input variables and hence 16 possible input combinations. But a large number of these are physically impossible because the platform can only be in one place at a time. Figure 7.24(*a*) illustrates the problem and shows the truth table which describes it. Plotting the Karnaugh maps for M, H and U in figure 7.24(*b*) leads immediately to the solutions. The network require two two-input AND gates, two two-input OR gates and some inverters. This is about a fifth of the logic that would be needed if the 1s in the truth table were realised directly (effectively assigning 0 to each X).

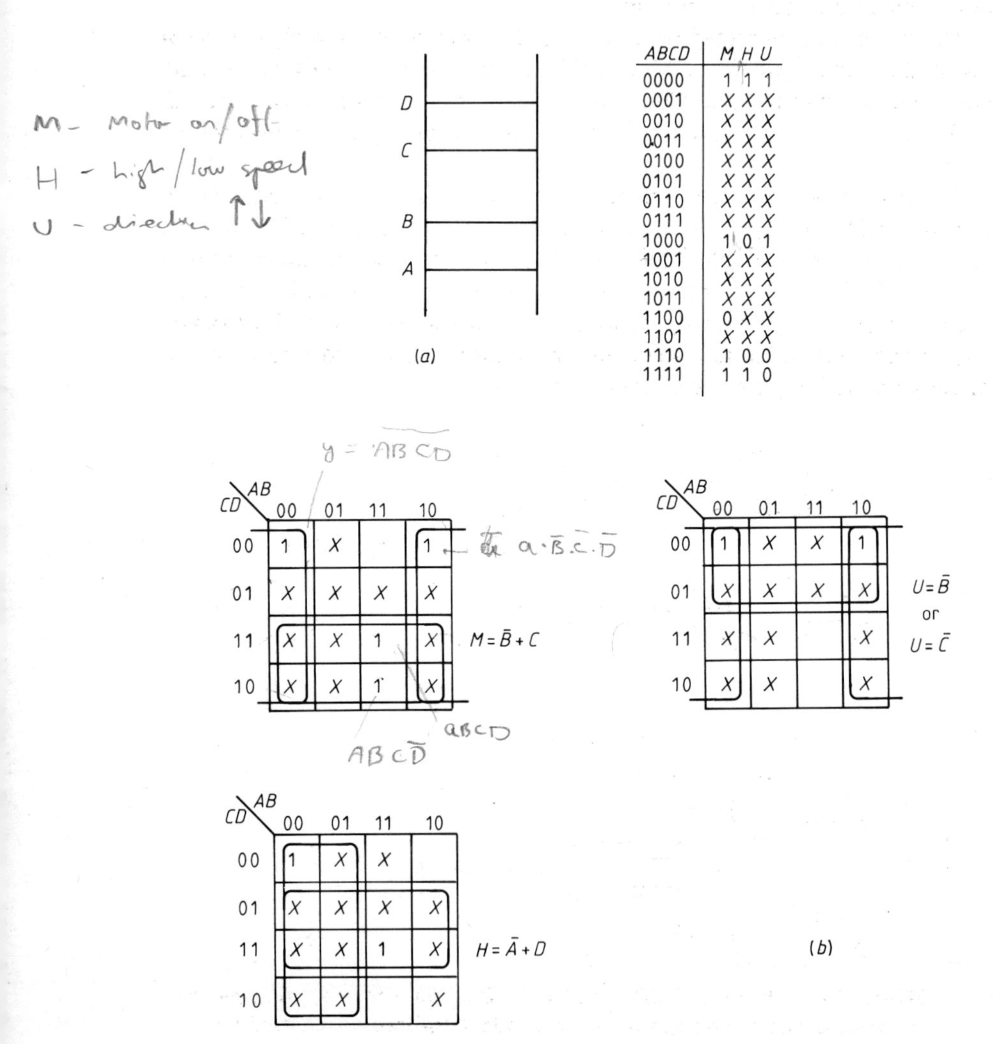

ABCD	M H U
0000	1 1 1
0001	X X X
0010	X X X
0011	X X X
0100	X X X
0101	X X X
0110	X X X
0111	X X X
1000	1 0 1
1001	X X X
1010	X X X
1011	X X X
1100	0 X X
1101	X X X
1110	1 0 0
1111	1 1 0

Figure 7.24(*a*) Schematic diagram and truth table for a platform in a vertical shaft. Sensors A, B, C and D generate a logic 1 when the platform is above their level. Output signal M turns the motor on and off. H selects high or low and U selects up or down; (*b*) maps for the motor control signals.

This dramatic reduction is a result of the freedom to choose output values for most of the input combinations. In this case, as in most extreme examples, the final expression agrees with the intuitive solution to the problem which could probably have been deduced without the need for truth tables or Karnaugh maps. The motor control signal, for example, must be ON when the level is below B ($B=0$) or above C ($C=1$). Nevertheless the formal approach assures a valid solution.

The expressions obtained from truth tables or from maps imply the use of AND and OR gates. However, it is possible to use NAND or NOR instead. A NAND gate followed by inversion yields the AND function. Inspection of the truth table for a NAND gate shows that inversion of each input provides the OR operation.

So an alternative realisation of a sum of products expression such as $F=AB+CD$ is possible with NAND gates and is shown in figure 7.25(*a*). The cascaded inverters are obviously redundant, so the network reduces to that in figure 7.25(*b*). The same result can be obtained algebraically. The final NAND network is topologically identical to the direct realisation with AND and OR, so the truth table and map design methods developed above can be directly applied to yield the all NAND solution.

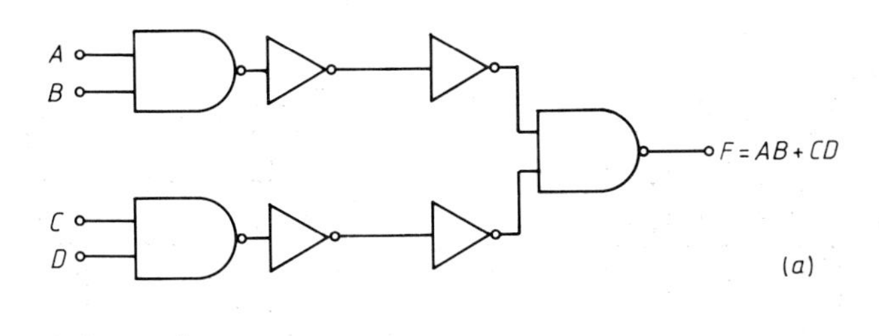

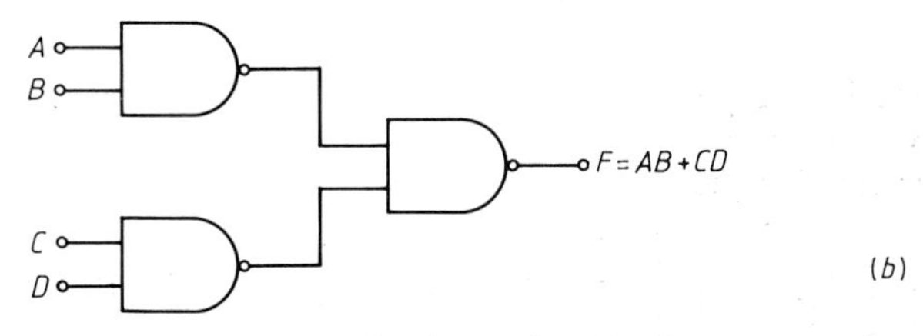

Figure 7.25 All NAND realisation of a Boolean sum of products expression. The inverters in (*a*) are redundant and can be removed to give (*b*).

It seems natural to produce realisations from truth tables or maps which yield the true output function. However, it is also possible to form the inverted function by realising 0s. The true output can then be obtained by inversion. Although this may need an extra logic operation, it sometimes yields a more economical solution.

A further variation is to form the output function by OR operations on the inputs followed by an AND operation. Manipulation of the algebraic sum of products expression using De Morgan's theorem yields this product of sums form. There is normally no advantage in this approach, and although it is entirely equivalent to the AND–OR method, it does not follow directly from the truth tables or maps and is rarely employed.

7.7 Sequential Logic, Memory and Bistables

Apart from the time delay inherent in all gates, the output from a combinational logic system is a function only of the current values of the input variables. Sequential logic provides an output which depends also on previous values of the input signals. Thus it has memory. Much more interesting signal processing then becomes possible.

It is possible to design and construct sequential logic from combinational logic with overall feedback. However, it is usually more convenient to use ready-made bistable elements which have internal feedback to produce memory functions.

The D bistable

One of the simplest bistables is shown in figure 7.26. It has an input D and a further control input, clock (Ck). The output Q maintains its state regardless of the value of D so long as the clock is at 0. When the clock is pulsed to 1 and back to 0, Q takes the value of D and retains it. There are minor differences between variants of the D bistable as to precisely at what point during the clock cycle the output changes.

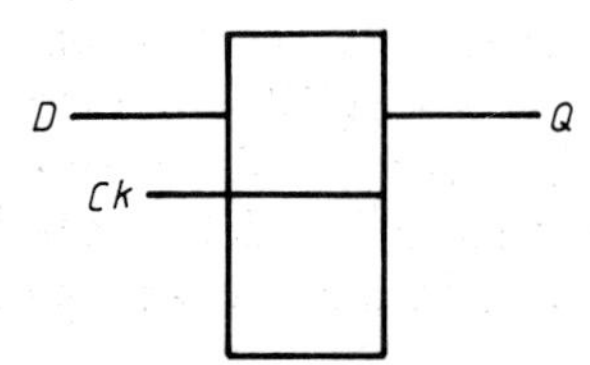

Figure 7.26 A D bistable.

Hence the D bistable is an information storage device. It can acquire and remember one bit of digital data until its contents are altered. The function of the D bistable can be summarised by the truth table in figure 7.27 which defines the output state before (Q_n) and after (Q_{n+1}) a clock pulse.

Practical D bistables often have extra connections. An inverted output $\bar{Q}$

D	Q_n	Q_{n+1}
0	0	0
0	1	0
1	0	1
1	1	1

Figure 7.27 Truth table for a D bistable before (Q_n) and after (Q_{n+1}) a clock pulse.

may be available, and inputs which force the output to 1 or 0 (PRESET and CLEAR), independent of the state of the clock, may be provided.

Registers

An array of D bistables can be used to store a binary word. Figure 7.28 shows four bistables arranged to hold a four-bit parallel word presented at the D inputs. All of the clocks are connected together so that a single clock pulse activates the register and stores the input word which then becomes available in parallel on the Q outputs.

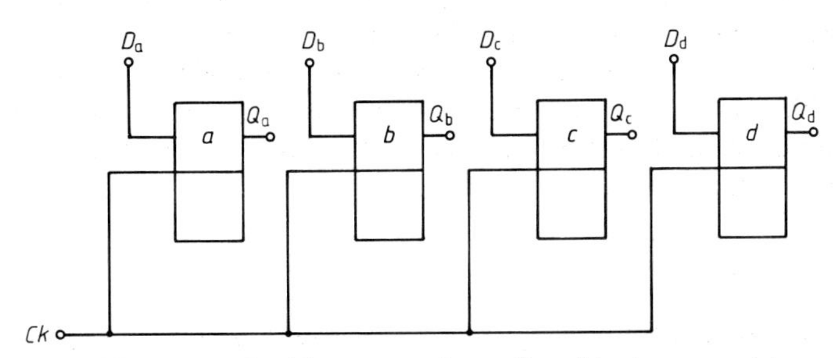

Figure 7.28 Four D bistables arranged as a four-bit storage register.

Serial data transmission to and from a register is possible if the bistables are arranged as a shift register. Figure 7.29(*a*) shows a four-bit shift register formed by connecting the Q output of one bistable to the D input of the next. The D input of the first bistable is the input to the shift register, and an output is available from the Q of the last element in the chain.

Suppose the register initially contains the word 1011 and the serial input D_a is held at 0. After the first clock pulse, bistable d will have taken the value from c. Bistable c will have taken the value of b. The data in a will have been transferred to b and a will be 0. Thus the information has been shifted one place to the right.

After four clock pulses, the information held in parallel in the register will have been presented at Q_d in serial form and replaced by new data (all 0 in this case) entering from the serial input. Figure 7.29(*b*) shows the contents of the register at each stage of the shifting process. Likewise, a stream of serial data arriving at the serial input synchronised to the clock pulse train can be entered into the register.

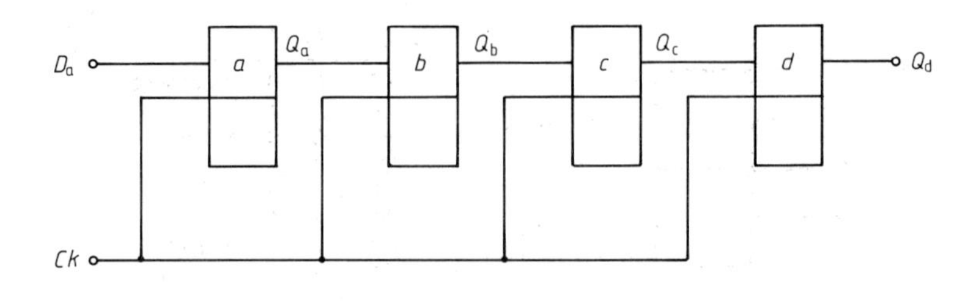

Clock pulse	Q_a	Q_b	Q_c	Q_d
0	1	0	1	1
1	0	1	0	1
2	0	0	1	0
3	0	0	0	1
4	0	0	0	0

(b)

Figure 7.29(*a*) Four *D* bistables arranged as a four-bit shift register; (*b*) the contents of the register as the clock is pulsed.

A parallel output word can be obtained from the *Q* outputs of the individual bistables. Data can be forced into the register in parallel if the bistables have individual PRESET inputs. Parallel access to the register obviously requires more external connections, and may be limited by the number of pins on a package.

A shift register with serial input and parallel output is a common method of achieving serial to parallel conversion of *n*-bit data. Serial data is presented at the shift register input and becomes available in parallel form after *n* clock pulses. The reverse process can be accomplished with a parallel-in serial-out register.

7.8 *JK* Bistables and Counting

More general sequential operation is produced by the *JK* bistable which is shown with its truth table in figure 7.30. Four actions are defined by the state of the two inputs, *J* and *K*. The first three input combinations provide data storage similar to that of the *D* bistable. But when *J* and *K* are both 1, the output state reverses (toggles).

The majority of *JK* bistables have a two-stage mode of operation. At the leading (positive-going) edge of the clock, pulse data is accepted from the *J* and *K* inputs and stored internally. Then at the trailing edge the output *Q*

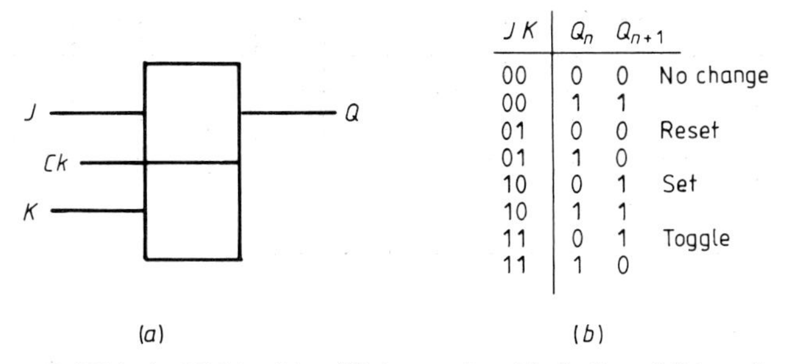

JK	Q_n	Q_{n+1}	
00	0	0	No change
00	1	1	
01	0	0	Reset
01	1	0	
10	0	1	Set
10	1	1	
11	0	1	Toggle
11	1	0	

Figure 7.30(*a*) A *JK* bistable; (*b*) its truth table before (Q_n) and after (Q_{n+1}) a clock pulse.

takes up the appropriate value. This master–slave action ensures an orderly transfer of data between cascaded bistables.

Frequency division and counting

The toggling mode of the *JK* bistable can be exploited to make counters. A single *JK* bistable with the *J* and *K* inputs held at 1 will change state for every complete input clock pulse. After two clock pulses the output will have returned to the original value. So the rate of the output pulse train is half the rate of the clock input. A chain of *n* cascaded *JK* bistables can divide a frequency by 2^n.

Pulse rate division is equivalent to counting. Figure 7.31 shows three cascaded *JK* bistables with their *J* and *K* inputs all held at 1. An input signal is fed to the clock input of the first bistable *C*. The output from *C* forms the clock input of *B* and the output of *B* drives *A*.

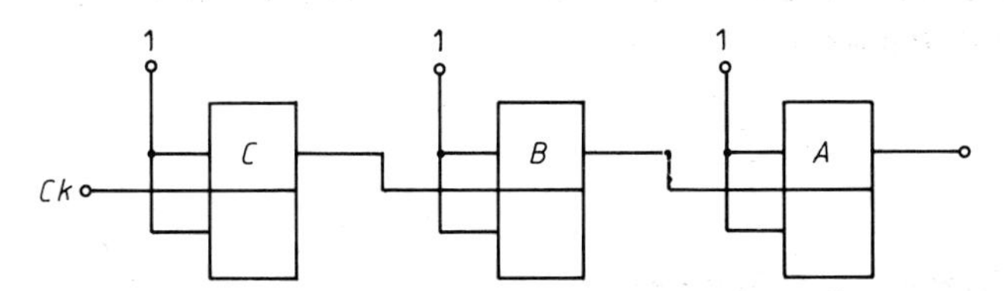

Figure 7.31 A three-stage binary divider.

Suppose *ABC* = 000 initially. After the first clock pulse, *C* will have toggled to 1. Bistable *B* has not yet received a complete clock pulse and remains at 0. After another clock pulse *C* returns to 0 completing the clock pulse to *B* which toggles to 1. The next clock pulse sets *C*. When *C* returns to 0 on the following pulse, *B* returns to 0 setting *A*. Inspection of figure 7.32 shows that *ABC* is counting up through a binary sequence. After eight clock pulses the counter contents *ABC* are reset to 000 and the cycle repeats.

Clock pulse	ABC
0	000
1	001
2	010
3	011
4	100
5	101
6	110
7	111
8	000

Figure 7.32 The contents of the three-stage divider as the clock is pulsed.

This type of counter has some limitations. It is restricted to counting in a pure binary sequence whose length is an integral power of two. Also, there must be delays as the result of each input pulse propagates down the chain. The bistables do not acquire their new states simultaneously and transient invalid codes can occur as the changes ripple through the counter.

Figure 7.33(*a*) shows how the asynchronous (ripple through) counter of figure 7.31 can be modified to terminate the sequence at a count of, say, five. The bistables must have *clear* inputs which force the Q outputs to 0 regardless of the states of J, K and the clock. The counter is reset to 000 when it reaches the state $AB\bar{C}$ and the map in figure 7.33(*b*) shows that this condition can be grouped with ABC (because ABC is never reached). So the required function is $R = AB$, just an AND operation.

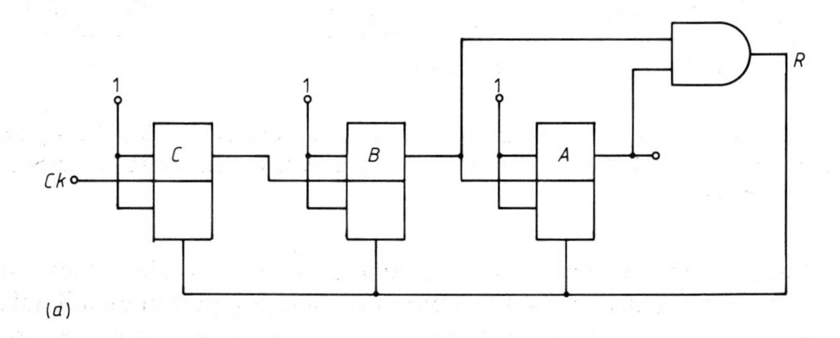

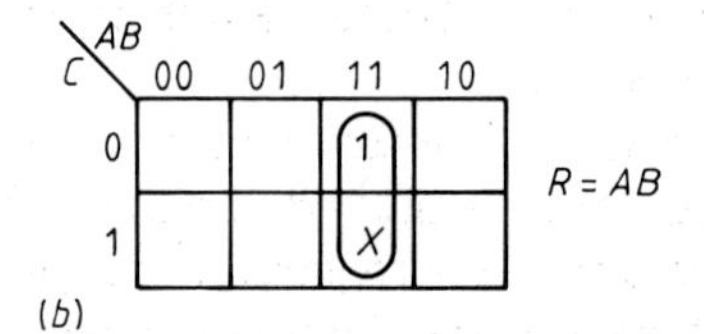

Figure 7.33(*a*) A five-state ripple through counter; (*b*) generation of the RESET pulse *R*.

Synchronous counting can be achieved by applying the clock signal simultaneously to all bistables in the counter. The state of each bistable after a clock pulse and thus the next count too can then be determined by setting up the values of J and K before the clock pulse. These values can be produced by combinational logic operating on the preceding count. The extra freedom allowed by this approach enables the design of counters in any arbitrary code. As an example we shall consider a counter which operates in a Gray code. The form of the counter and the count sequence are shown in figure 7.34(*a*).

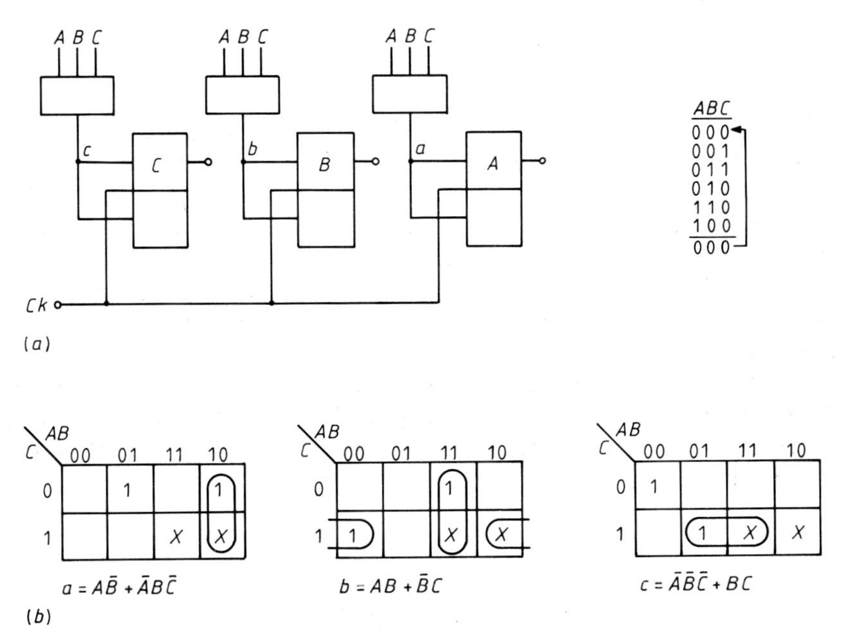

Figure 7.34(*a*) A synchronous three-bit counter in Gray code using the toggling property of the *JK* bistable; (*b*) maps for the generation of the *JK* inputs from the counter states.

There are six different states of this counter, so three bistables at least are needed. For simplicity we shall connect the *J* and *K* inputs of each bistable together. This restricts us to using only two of the four possible input conditions. For $J=K=0$ no change occurs whilst the state reverses for $J=K=1$.

Each of the combinational logic blocks forming *a*, *b* and *c* will have the register contents *ABC* as inputs. For every combination of *A*, *B* and *C* where the bistable is required to toggle, *J* and *K* must be 1, otherwise they are 0. There are two CAN'T-HAPPEN conditions, $ABC=101$ and $ABC=111$.

The maps for *a*, *b* and *c* are shown in figure 7.34(*b*). In all cases the CAN'T-HAPPEN conditions allow some simplification and more design freedom can also be allowed by separating the *J* and *K* inputs. This introduces DON'T-CARE conditions because the bistables can be set and reset as well as toggled. More *X* cells lead to greater simplification but six logic blocks are then needed.

7.9 Arithmetic

Counting is an arithmetic operation and it is easy to see that a counter could be made to implement some form of addition or subtraction of pulses. A more general form of arithmetic operates on data presented as binary words either in parallel or in serial form.

Addition is the basic arithmetic operation. Figure 7.35 shows two four-bit binary words A and B which are to be added. The least significant bits of A and B must be combined to produce the least significant bit of the sum output. If both bits are 1 then the sum is 0 and a carry is generated to be added into the next most significant pair of bits. The addition of higher order pairs of bits must make provision for including any carry from a lower stage. The rules for the binary addition of two bits are very simple and are shown in the truth table of figure 7.36.

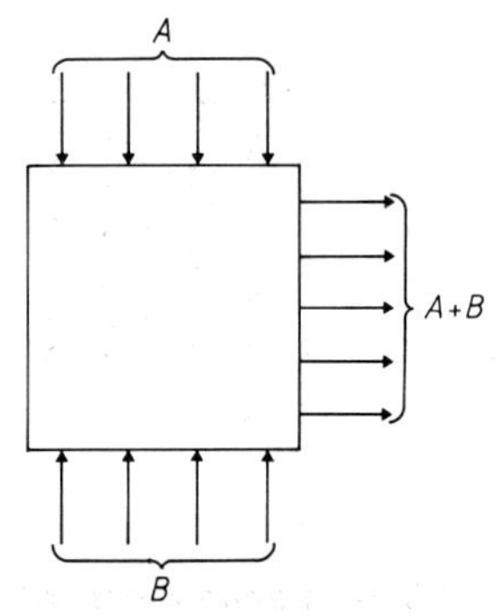

Figure 7.35 Addition of two binary words A and B to provide the sum $A+B$.

A_n	B_n	S_n	C_n
0	0	0	0
0	1	1	0
1	0	1	0
1	1	0	1

Figure 7.36 Truth table for the half adder combining A_n and B_n to give the sum S_n and carry forward C_n.

We recognise S as the EXCLUSIVE OR function and C as the AND function. The corresponding gate network is shown in figure 7.37(*a*). Alternatively, if we choose to realise $\bar{S}$ and then invert it, the two networks can be combined as shown in figure 7.37(*b*).

The combinational logic which produces the sum and carry from the addition of two bits is a half adder. A full adder is needed for the higher order bits, which must add in the carry from the next lower order. The full adder can be realised from half adders or it can be formed directly from its truth table shown in figure 7.38.

Subtraction can be performed in a similar manner, using gates to implement the rules for binary subtraction. However, it is usually more convenient to adopt a representation of negative numbers which allows subtraction to be achieved with the same hardware as addition.

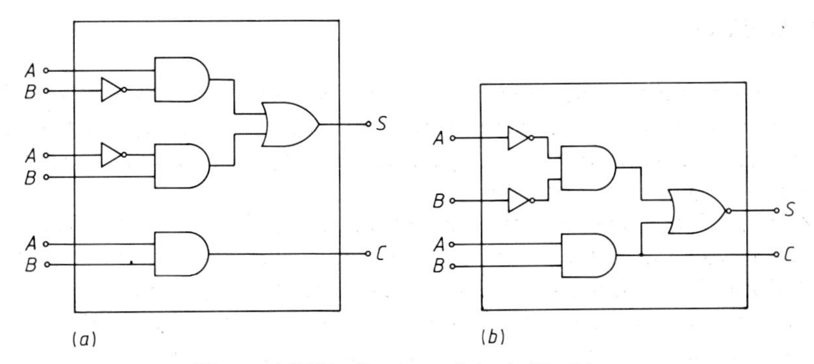

Figure 7.37 Realisation of the half adder.

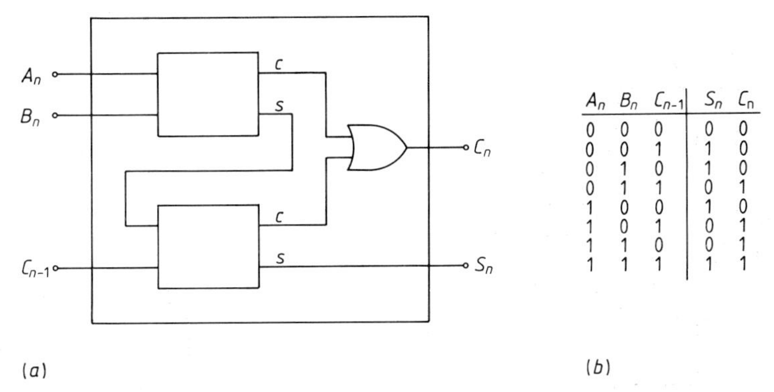

A_n	B_n	C_{n-1}	S_n	C_n
0	0	0	0	0
0	0	1	1	0
0	1	0	1	0
0	1	1	0	1
1	0	0	1	0
1	0	1	0	1
1	1	0	0	1
1	1	1	1	1

Figure 7.38 The full addition of two bits A_n and B_n with a carry forward from the previous stage C_{n-1}.

Negative numbers

Both the pure binary code and BCD considered in §7.3 represent only positive integers. Pure binary can readily be extended to negative integers by including a bit which indicates the sign of the number. The usual convention is to position this sign bit at the extreme left of the word. Then if the remainder of the word represents the magnitude of the number we have a notation analogous to that used in conventional decimal arithmetic. Thus $+5$ could be represented by 0101 and -5 by 1101.

Signed magnitude representation is 'natural' but not as convenient as a complement form. Two's complement notation represents a negative binary number as the difference between that number and the power of two equal to the number range. So in a four-bit system $+5$ is still 0101 but -5 becomes $10000-0101=1011$. Table 7.2 shows the signed magnitude and two's complement representations of numbers around zero in a four-bit system.

Table 7.2 Signed magnitude and two's complement representation of numbers around zero in a four-bit system.

Decimal	Signed magnitude	Two's complement
+3	0011	0011
+2	0010	0010
+1	0001	0001
+0	0000	0000
−1	1001	1111
−2	1010	1110
−3	1011	1101
−4	1100	1100

If overflow beyond four-bits is neglected, it will be seen that numbers in two's complement notation are continuous through zero. Then the hardware just developed for adding positive numbers will work with negative numbers too. Thus, for example 1110 (−2) + 0100 (+4) = 0010 (+2), where the sign bit has been treated as the most significant bit of an ordinary binary integer and the overflow (carry) has been neglected. Another example could be 0010 (+2) + 1101 (−3) = 1111 (−1). Adding −3 is equivalent to subtracting +3 so it is unnecessary to provide special subtraction hardware.

Nevertheless it may still be necessary to form the two's complement of a binary number. Fortunately there is no need to do the explicit subtraction implied by the definition of the notation above. A more convenient route is to invert all the bits of a word, forming the one's complement, and then to add 1 (see figure 7.39).

+5	0101
One's complement	1010
add 1	1
Two's complement	1011

Figure 7.39 Generation of two's complement number through the one's complement.

Arithmetic is possible on BCD encoded numbers. The operations are a hybrid of normal decimal manipulation with binary arithmetic on the four-bit numbers. Figure 7.40 shows a simple example of addition of two BCD numbers. Note that if the addition result of two BCD digits is greater than 9, a valid BCD number is formed by adding 6 and generating a carry forward to the next stage. Negative numbers can be represented using a ten's complement form of BCD. Most calculators hold their data in a BCD form and hence use BCD arithmetic.

```
  257   0010   0101   0111
+ 124   0001   0010   0100
        ----   ----   ----
        0011   0111   1011
                  1←┐ 0110
  ---   ----   ----  ----
= 381   0011   1000 (1)0001
```

Figure 7.40 Addition of BCD encoded numbers. After binary addition of each digit, 6 is added if the sum exceeds 9 to generate the proper BCD digit.

Multiplication and division

Multiplication of binary integers can be accomplished by successive addition and shifting in a manner similar to that used for decimal long multiplication. The final product is formed as the sum of the partial products of the multiplicand with the bits of multiplier. In binary arithmetic the multiplier bits can only be 0 or 1 so the partial products are just zero or the shifted multiplicand itself. Figure 7.41 shows the mechanism for a four-bit by four-bit problem. However, it is usually more convenient to accumulate the partial products in a single register rather than provide several registers and sum their contents at the end as shown in this example.

```
   5          0101
× 11          1011
              ----
              0101
             0101
            0000
           0101
           -------
= 55       0110111
```

Figure 7.41 Multiplication of two binary integers by repeated shift and add operations.

Direct multiplication by combinational logic is also possible and is much faster than the shift and add process. Unfortunately the amount of logic needed increases rapidly with the word length; ready-made multipliers integrated onto a single chip are available. These are extremely fast, producing the product in a few gate delay times, but are very expensive and power thirsty.

Division is more complicated than multiplication, but can be achieved by a process analogous to decimal long division involving repeated subtraction and shifting.

Arithmetic operations on numbers in representations other than the signed and unsigned integer forms of binary and BCD considered here are possible, but they require extensive circuitry.

8

Large Scale Integration

The rapid advances in semiconductor technology have led to the introduction of large scale integrated (LSI) circuits containing many thousands of transistors. LSI circuit technology has been exploited in several ways, including custom circuits specifically designed to provide a particular signal processing or control function, and uncommitted logic arrays (ULA) which can subsequently be interconnected to fulfil a specific application. Two other areas where the results have been particularly spectacular are in the fabrication of memory elements and microprocessors.

8.1 Large Scale Memory

There is an obvious requirement for data storage in digital systems. However, relatively small amounts of data can be held in bistable registers of the type discussed in §7.7. Parallel data storage in registers gives very rapid access, but the number of input and output connections required severely limits the total quantity of data that can be held in a single integrated circuit.

Serial shift register storage needs only a pair of input and output connections regardless of the length of the register and has useful applications where the data is to be accessed in the order in which it is stored. However, access to random locations within the shift register requires that the whole data stream be shifted until the desired item is at the output. If the register can only be shifted one way it will be necessary to wait on average for half of the data to be shifted out before the item is reached. As the register length increases so does the average delay.

A more generally useful form of large scale memory is organised as an array of parallel registers of which only one may be accessed at a time. The particular register is selected from the array by an address presented to the memory. In most cases a single set of connections serves for both writing data to and reading data from the selected location. A control line determines the direction of data flow along the bus and the appropriate internal function.

In order to achieve a very high density of data storage in practice the individual memory elements are not conventional bistables but are made with more economic circuitry, sometimes using only a single transistor. In this way it is possible to store several hundred thousand bits of information on a single integrated circuit chip.

A possible structure for an LSI memory is shown in block diagram form in figure 8.1. The total storage capacity is 64K bits organised as 8K × eight-bit words where K is a multiplier of $2^{10} = 1024$. A 13-bit address code is needed to identify the current location within the array and an eight-bit bi-directional data bus allows the transfer of words to and from the selected location. In addition to the read/write line, a chip enable line is provided. When the chip is disabled, the tristate output buffers are allowed to float so that other devices connected to it may transfer information.

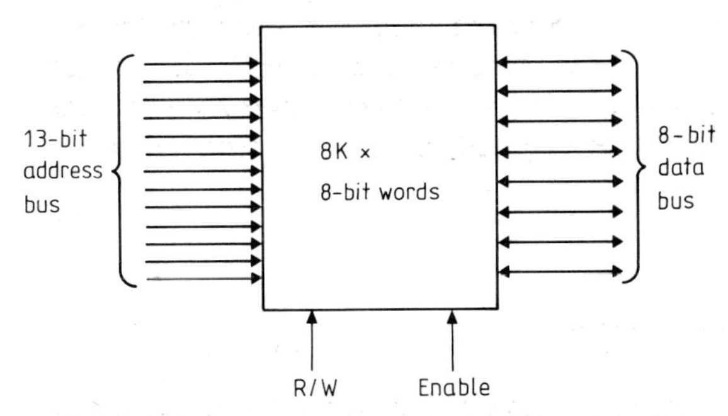

Figure 8.1 A 64K LSI memory, organised as 8K × eight-bit words.

Memory organisation in the form of eight-bit words (bytes) is convenient for small systems. Chips intended for use in large systems usually hold only one-bit words. An array of n such memories with their address and control lines connected in parallel can then be used to store n-bit words. This arrangement reduces the number of pins required on the memories and hence reduces costs.

LSI memory which allows writing and reading data to and from any randomly selected location is called random access memory (RAM). Two types of RAM are available: static memory behaves like bistable storage and retains data written to it as long as power is supplied or until rewritten; dynamic memory, which relies on charge storage on the gate of a MOST, requires refreshing every few milliseconds as the charge leaks away. In older dynamic memories some external circuitry was needed to perform the refresh operation but more modern devices have internal refresh and the user need hardly be aware of the process. Dynamic memory cells are very compact, so dynamic memories are denser than static ones and consequently store more data per chip.

The uses of RAM are obvious, but it may not be apparent why memory which contains permanently written data and can only be read is also in common use. Read only memory (ROM) does have, however, many applications.

8.2 Applications of ROM in Logic

All of the combinational and sequential logic functions discussed in the preceding chapter can be implemented with LSI memory elements rather than gates or bistables. A bistable memory element is a relatively complicated assembly of gates, but the storage cells in LSI memory, particularly in ROM, can be very simple. A high degree of regularity leads to very compact layouts and low unit costs. Memory realisations of logic can be both convenient and economical.

Any combinational logic function defined in a truth table or in the expanded sum of products form can be directly translated into a ROM implementation. The input variables form an address which points to a location where the corresponding output signals are stored. So the ROM becomes just a set of look-up tables describing the combinational logic function. The size of ROM needed will be $n(2^m)$ bits, where m and n are the numbers of input and output variables respectively. There can be no simplification as is possible with gate realisations even if CAN'T-HAPPEN or DON'T-CARE input conditions are present. This means that for functions which can be reduced to very small gate networks the ROM implement is unlikely to be efficient. But problems with either little simplification or a large number of variables will be good candidates for the ROM approach.

A common problem which is ideally suited to ROM solution is that of code conversion. For example the conversion between the four-bit pure binary and XS3 codes shown in figure 8.2 requires a ROM containing 16 four-bit words (XS3 is a code sometimes used instead of pure BCD to represent the decimal digits 0–9). This is a very small ROM and the one-chip memory solution can be compared with the gate implementation which needs about 20 gates and inverters. The design time is just that required to write down the truth table describing the input and output codes.

A further advantage of the ROM approach is that the function can easily be modified. A minor change in the behaviour of a system might require the complete redesign of a gate version whereas the ROM approach may need only a change in the contents of the ROM.

The first ROM to be produced had their stored data determined at the time of manufacture. This has the disadvantage that it is only economical if very

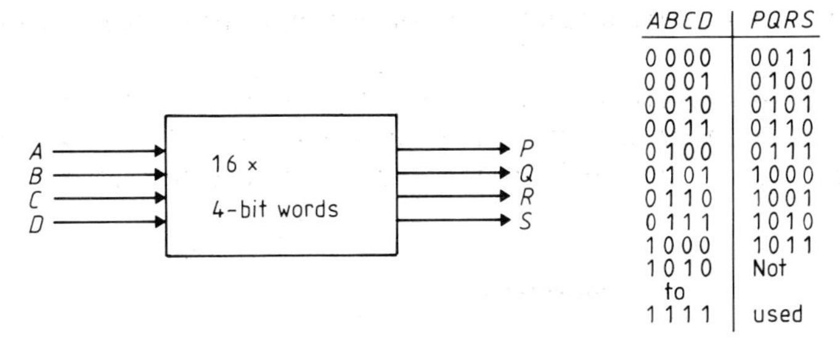

ABCD	PQRS
0000	0011
0001	0100
0010	0101
0011	0110
0100	0111
0101	1000
0110	1001
0111	1010
1000	1011
1010 to 1111	Not used

Figure 8.2 Binary to XS3 code conversion with a ROM.

large numbers of identical units are produced. There was an obvious need for ROM which users could tailor to a particular application by writing their own data into the memory. Then the ROM approach would become suitable, even when only a few units were needed.

Programmable ROM (PROM) can be written to once only and thereafter retains the data permanently. More recently, PROM has been introduced which can be erased either by exposure to ultra violet radiation (EPROM) or by applying an electrical signal (EAROM). This type of memory has both read and write capabilities but the writing and erasing process is very slow compared with the writing of data into an RAM. EPROM may take many minutes to erase and rewrite whereas the contents of RAM can be changed in tens or hundreds of microseconds. Moreover, the storage properties of erasable PROM become degraded after a few hundred 'erase and rewrite' cycles.

8.3 Sequential Logic

If feedback is applied to ROM so that the input address depends on the output data, it becomes possible to implement sequential logic functions. The general arrangement is shown in figure 8.3 where some of the output signals can be fed back to form part of the ROM address. Registers are normally required at the input to synchronise changes to the clock signal in the system. The output signals are decoded by a second ROM to form the final outputs.

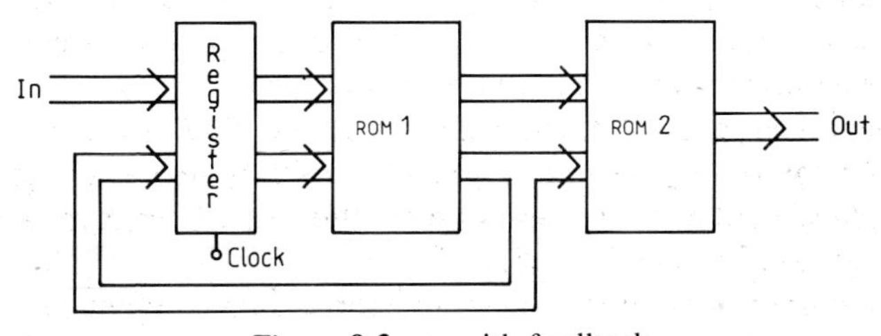

Figure 8.3 ROM with feedback.

At each clock pulse a new address formed from the current input signals and the output data from the last clock pulse is presented to the ROM. The ROM then produces new output data ready for the next clock cycle. So every time the clock is pulsed, the system takes up a new state which is dependent on the history of the system.

Figure 8.4 shows a three-bit counter in Gray code which is a simple example of a sequential circuit. There are six valid codes in this count sequence and each code must be uniquely distinguished by a combination of the input address. The lowest power of two greater than six is eight, (2^3), and so we require three address variables. As there is no input signal to the counter other than the clock, there will then be three feedback lines.

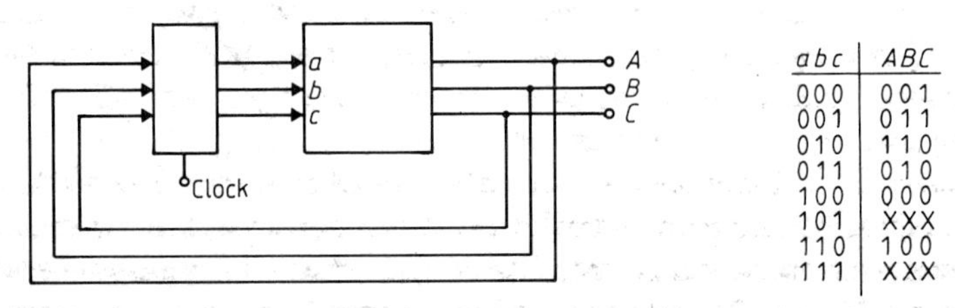

abc	ABC
000	001
001	011
010	110
011	010
100	000
101	XXX
110	100
111	XXX

Figure 8.4 ROM implementation of a Gray code counter.

If we choose that the output code *ABC* is taken directly from the ROM so that no further encoding is required, then *ABC* at any time defines the input address *abc* for the next clock pulse. Thus it is clear that each location of the ROM must contain data which forms the subsequent code in the count sequence. (Location 000 must contain 001, 001 must contain 011 etc.)

There are two CAN'T-HAPPEN conditions, so in principle the corresponding ROM locations will never be accessed and the data contained there can be arbitrary. However, if in some unexpected circumstances a supposed CAN'T-HAPPEN condition does arise, perhaps at power-up, it would be sensible to program these ROM locations with data which will force the counter back into the correct sequence. 000 would be a suitable choice, resetting the counter back to the beginning of the sequence.

Sequential systems of this kind with a finite number of internal states are called finite state machines. They are widely used in their own right and as components of larger systems such as the processors considered in §8.4.

If the number of input and feedback signals becomes large, the size of the ROM needed increases very rapidly. In most cases a high proportion of the possible states cannot occur and so a corresponding part of the ROM is effectively wasted. Only a small part of the ROM contains useful data and the ROM solution has become inefficient. A better solution in such cases is to use a programmed logic array (PLA).

A PLA contains an array of AND gates whose outputs are combined with OR gates to form the final outputs. Programming selects the terms in the expanded 'sum of products' Boolean expression defining the functions to be implemented. At present PLA are mostly used as subsystems within larger LSI circuits. However, ready programmed and field programmable logic arrays are themselves available.

Custom LSI and ULA

The full potential of IC technology can be realised if a circuit, either analogue or digital, containing transistors and other components is specifically designed to perform the required function. The design effort needed to produce such a circuit is very large and is only justified when the cost can be spread over a large number of identical units. Custom LSI is not economical until the production quantities reach many thousands—preferably much higher—or unless the application is so important that the very high development costs can be tolerated.

Uncommitted logic arrays (ULA) offer an alternative solution for many instrumentation and control problems. A ULA consists of an array of circuit elements—bistables, gates, transistors and other components—integrated onto a single chip. No interconnections are made between the components until the application is known. Then a metallisation pattern can be designed to interconnect the components and implement the required function. It is the interconnection pattern that tailors the general purpose chip, which can be produced in quantity, to a specific application. A further advantage of the ULA approach is that the design time is substantially reduced compared with a custom LSI circuit.

8.4 The Microprocessor

The concept of tailoring general purpose components such as ROM or ULA to suit specific applications is carried further with the microprocessor. A microprocessor is a digital system integrated onto a single chip. This is capable of performing relatively simple arithmetic and logical operations, together with some other data handling functions.

In principle any digital process, no matter how complicated, can be built up from a sequence of simpler operations. Only a few fundamental operations are absolutely necessary. However, it is more convenient if more than the minimum number are available and most microprocessors have a repertoire of about a hundred different operations.

Binary codes presented to the microprocessor define the series of operations to be performed. A program is the sequence of these instructions required to carry out the process. The program is stored in memory and further memory may be needed to hold data during the execution of the program.

A system comprising a microprocessor, memory and circuitry for buffering input and output signals constitutes a microcomputer. Sometimes this term is reserved for the case where all these components are integrated on a single chip. Usually though, it is understood to mean any computer system using a microprocessor. Because a microprocessor can be used for a wide range of different applications by appropriate programming, very large numbers of the same type can be produced and they then become very cheap.

The availability of cheap processing power has been exploited in two ways. At one extreme computing systems based on microprocessors have become common. Desktop and personal computers are now widely used in industry, commerce and education, augmenting or replacing minicomputers or computing facilities provided by large main-frame computers. In many cases the low cost of a microcomputer allows computational techniques to be employed where they were previously uneconomic.

Although this may be the most obvious impact of the microprocessor, it is not the area in which microprocessors are most widely used. Many millions of microprocessor chips have been produced since their introduction in 1973. The great majority of these have been incorporated into instruments or machines for signal processing and providing control functions. In some cases the application can be very mundane indeed, perhaps just controlling the speed of an electric motor or monitoring a set of switches. Similar control functions can be provided by conventional analogue or digital electronics, but the microprocessor solution may be either cheaper or more flexible. Programmability allows change or enhancement of function at a later date without modification of the electronic components. At a rather higher level, a microprocessor might be incorporated into an electronic instrument to provide some processing of data, or perhaps to provide superior and more convenient operation of the instrument for the user. Microprocessors have become so cheap that several can be incorporated into a single system, either working independently on separate tasks, or cooperating on the same task.

8.5 Microprocessor Structure and Organisation

Although there are considerable differences in detail, most current microprocessors have an internal structure similar to that shown in figure 8.5. An arithmetic and logical unit (ALU) is connected to a number of registers, counters and buffers through one or more parallel data buses. The first microprocessor marketed operated on four-bit data words and similar devices are still useful for very simple control and relatively limited operations on BCD data, such as required in calculators. However, the eight-bit devices introduced a little later were suitable for a much wider range of applications and have since become widely used. More recently, powerful 16-bit and 32-bit machines have also become available.

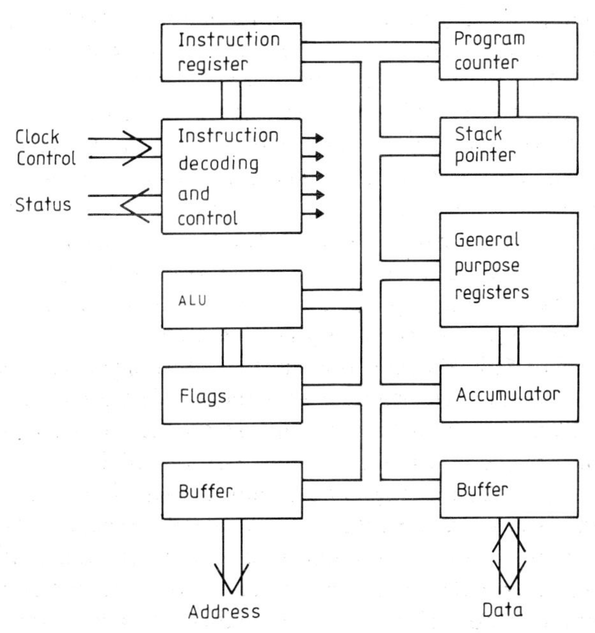

Figure 8.5 Structure of a typical small microprocessor.

Address and data buses allow the processor to exchange data with memory. Although eight-bit processors operate on eight-bit data words, they usually have a 16-bit address formed by concatenating two eight-bit words so that they can directly access $2^{16} = 64\text{K} = 65536$ memory locations. Memory is now sufficiently cheap that 16-bit and 32-bit processors have even wider address buses and can directly address very large memory spaces.

Every operation that the processor can perform consists of a sequence of even more elementary data manipulations within the processor structure. These micro-orders are functions, such as incrementing or decrementing data in a register, shifting data left or right or transferring data between a register and a buffer.

At the beginning of program execution, the program counter contains the address of the first instruction of the program. This address is transferred to the address buffer and presented to the memory. The operation code defining the instruction is then read and fetched into the instruction register and the program counter is incremented to point to the next location. Logic in the instruction decoding block produces the sequence of control signals required to carry out the micro-orders which implement the instruction.

Addressing modes

Some instructions operate on data already held within the processor, perhaps in one of the general purpose (scratch-pad) registers. However, most instructions which process data require access to operands held in memory and must specify an address. In immediate addressing, the operand is stored immediately after the operation code as a part of the program. So the program counter will already be pointing to the required data, and one or more memory read and program counter increment cycles will obtain the operand.

This addressing mode is straightforward but requires that the data is fixed and known at the time of writing the program. If the program is stored in ROM, the data cannot be changed. Even if the program is in RAM, it is bad practice to modify the program during execution.

A more general scheme for accessing operands is direct addressing, where the program contains an address pointing to data held in RAM rather than holding the data itself as in immediate addressing. In direct addressing, the address following the operation code is read, fetched into the processor and stored in the address buffer. Then, a memory read or write operation transfers an operand between the processor and the memory.

Direct addressing is inefficient when repeatedly accessing the same location or when accessing a sequence of adjacent locations. Some processors have an indexed addressing mode in which a base address is held in an index resister within the processor. An effective operand address is formed from the base address and an offset. Fewer program memory reads are required and indexed addressing is a powerful method of processing lists or tables. Indirect addressing adds yet more power and flexibility in specifying an operand address by using the contents of the location generated in direct or indexed addressing as a further pointer to the final operand location.

The stack pointer

A feature of many microprocessor structures is the provision of a last-in first-out (LIFO) stack. Data can be successively pushed onto the top of the stack, forcing all other stack contents down one further level. When a data item is pulled off the top of the stack the remaining data all moves up one level. At all times the last data item entered is available on the top of the stack.

Most microprocessors maintain their stack in external RAM and use a stack pointer register to hold the address of the top of the stack. As the stack grows or shrinks the pointer is adjusted to keep track and there is no need to shift all the stack contents.

8.6 Machine Instructions

The ALU performs arithmetic and logical operations on data held in the processor or in memory. Some operations require only one operand, for example, the incrementing of a counter or the inversion of all the bits in a register. Other operations may specify two operands and it is usual for one of these to be held in a special register, the accumulator, and for the result to be returned to the accumulator.

Arithmetic operations such as addition and subtraction, incrementing and decrementing are usually performed on two's complement integers. Some very simple processors have only limited arithmetic capability, perhaps just addition. Many eight-bit processors include some limited arithmetic operations on double length (16-bit) integers. Alternatively, multiple precision arithmetic can be performed by repeated single precision operations. Integer multiplication and division can be achieved by software implementations of the shift and add method mentioned in §7.9. More powerful processors may include multiplication and division of signed integers in their instruction set.

Integer arithmetic is quite adequate for many control and instrumentation applications. Floating point arithmetic needed for scientific calculations can be provided either by software routines using the basic integer operations, or by special arithmetic LSI chips which can be interfaced to a microprocessor, relieving it of the computational burden. Floating point hardware will be included in many microprocessors as IC technology continues to improve.

Arithmetic instructions treat binary words as numbers, but logical instructions operate on the individual bits separately. The OR operation, for example, on the words 01011000 and 01000001 gives 01011001, where each pair of bits in the two operands has been combined with the OR function to form the result.

The state of the ALU after an operation is indicated by flags. Some processors only have flags indicating that the result was zero or that overflow (carry) occurred. Others may include flags indicating the results of BCD or two's complement arithmetic. Often the flags are grouped together and treated as a register; they themselves may be operated on by logical instructions.

In addition to the arithmetic and logical operations which provide the computational power of a microprocessor, other instructions are needed. There must be instructions to move data between processor registers, and ones to communicate with external memory and peripheral devices. In many processors, the last two functions are combined by treating peripherals as memory locations. Mapping peripherals into memory space slightly reduces the amount of real memory a processor can have for program and data storage, but simplifies the structure and reduces the number of different instruction types. Other processors provide explicit input and output instructions for communication with peripheral devices.

Finally it is necessary to have instructions which affect program flow. Jump instructions place a new address into the program counter and hence force execution to be diverted elsewhere in the program. Branch instructions are similar except that the operand specified is an offset to be added to the old contents of the program counter to form the new address. Using offsets from the current program counter contents rather than absolute addresses has the advantage that the resulting code can be position independent. Programs written entirely in position independent code can be located anywhere within memory space and will run without alteration.

Conditional jumps and branches provide a mechanism for following alternative courses depending on the state of the condition flags. These have been set as a result of some previous arithmetic or logical operation and, on encountering a conditional instruction, the specified flag or combination of flags is tested. If the result of the test is true the diversion is taken, otherwise the instruction is skipped and execution continues with the next in sequence.

Subroutines

Most programs contain segments of identical code that are used many times. It is convenient and improves reliability if only a single copy of the routine is held in the program. This subroutine is used repeatedly as necessary. A jump or branch to the beginning of the code diverts program flow to execute the code. At the end of the subroutine, execution returns to the point where it left in the main program. It is necessary to have a mechanism for storing the current contents of the program counter when a subroutine is called and restoring them at the end.

A stack is an ideal way of storing the return addresses for subroutines. The contents of the program counter at the time of suspension of the main program can be saved on the stack and replaced by the subroutine starting address. At the end of the subroutine the old program counter contents are pulled off the stack and restored so that program execution continues where it left off. A subroutine may call another subroutine and the depth to which they are nested is limited only by the available stack storage space.

In addition to holding program counter return addresses, a stack can store register contents and other information when a subroutine is encountered. It is also useful as a general purpose temporary data storage area with the LIFO structure automatically returning data in the reverse order from which it was presented to the stack.

8.7 Interaction Between Hardware and Software

A general purpose microprocessor programmed to do a particular task will always be much slower than a digital system specially designed to perform only that function, assuming that similar hardware technology is used in both cases. Many instructions, each taking a microsecond or so, are required to implement an operation which could be directly produced by dedicated hardware in perhaps a few hundred nanoseconds. The speed disadvantage compared with pure hardware will vary widely, but is likely to be a factor of between a thousand and a million. This is the price to be paid for the flexibility of the programmed approach.

Sometimes, as with general purpose computation, flexibility is essential, or the task may be so complicated that special purpose hardware would be impracticable. At the other extreme, very high speed processing of digital signals demands dedicated hardware. However, for many control and instrumentation applications where microprocessors are likely to be used, there will be a choice between hardware and software. More likely, it will be sensible to consider a combination of the two.

As an illustration let us consider providing a counting function similar to the Gray code example considered earlier in §8.3. The hardware form would take about 10 SSI packages and cost a few pounds. The alternative software approach needs a program to accept input pulses, maintain a count and deliver the output code. This program might amount to about a hundred binary words of machine code defining the instructions and cost a few pence in storage requirements.

The software solution still needs hardware consisting of the microprocessor, its associated memory and input and output circuitry. This is likely to cost as much as the hardware solution, and it would probably be uneconomic to consider using a microprocessor to provide the counting function alone. Yet if the microprocessor hardware already exists for some other purpose, the marginal cost of the counter is just the cost of the extra program development and storage. However, a maximum counting speed of a few kilohertz is likely to be the best that the microprocessor could achieve compared with tens of megahertz for a hardware counter.

If we decide that a programmable solution is appropriate there may be other trade-offs between hardware and software to consider. Suppose the input to the counter comes from a mechanical switch. Almost all mechanical switches suffer from contact bounce where the initial closure is followed by a series of breaks and makes lasting a few milliseconds until the switch settles down. Thus a train of pulses is produced instead of a single pulse as intended. This will give a false count and debouncing of the switch signal is necessary.

Sometimes it is possible to suppress the false pulses with a capacitor shunted across the signal line, but a neater solution in logic systems is to use a two-way switch and the set and reset functions of a bistable. When the switch operates, the bistable changes state and retains that state despite contact bounce, provided that the bounce is not so severe that the switch makes contact in the opposite sense.

An alternative software approach might be to introduce sufficient delay after a switch change has been detected to ensure that contact bounce has finished before the program looks for the next input pulse. Ten milliseconds or so should be adequate in most cases. A slightly more sophisticated method would be to interrogate the switch at intervals of a few milliseconds, and only accept the input when a constant signal is detected. In either case only a few bytes of program storage are required at very little cost, but the penalty is a reduction in maximum counting speed.

8.8 Program Development at Low and High Level

It is important to realise that good design procedures are as necessary in program writing as they are in the production of reliable circuitry. Small programs can be, and very often are, produced in an empirical and intuitive fashion. Such programs may appear to perform the desired operation, but it is difficult to be confident that they will not malfunction. Large programs developed in this way often either fail to work at all or exhibit errors (bugs) which may be impossible to eradicate. Often tracing and removing one bug simply reveals or introduces others, further compounding the problem.

Regardless of the design aids that may be available to the program writer, the first and most important requirement is a clear and unambiguous description of what the program is to achieve. As with all design, it may prove necessary to modify the objectives later, but they must be defined initially. The next stage is to find an algorithm, a sequence of computable steps which will guarantee a solution. This may not be straightforward, particularly in control problems involving interaction with many external signals, as here the program flow may become very convoluted.

Flow diagrams are a convenient way of illustrating the operation of programs or segments of programs, especially the decisions to be made and the actions to be taken as a result. Other methods of representing program flow are also popular, but flow diagrams are widely accepted as a useful aid to logical program design. As an example of their use we shall consider the development of a simple machine code program for a microprocessor which interrogates a keypad, waits until a key has been pressed, and returns a value indicating which key it is.

We shall assume that the 32-key pad is encoded as a 8×4 matrix with 12 lines as shown in figure 8.6. This coding arrangement is commonly adopted to reduce the nmber of lines and the amount of interface circuitry needed at the expense of subsequent decoding. Resistors from the X input lines to the common line define the logic levels at 0 when no contact is made. When a key

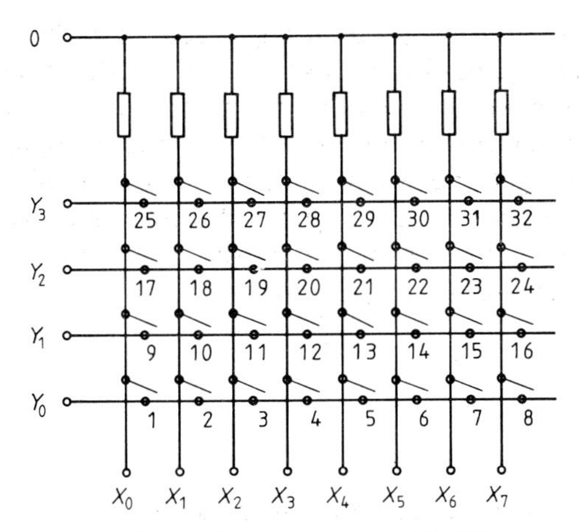

Figure 8.6 32-key pad encoded as a matrix of 8 *X* lines and 4 *Y* lines.

is pressed contact is made between one of the *X* lines and one of the *Y* lines. If we present signals to the 4 *Y* lines, a scan of the signals present on the *X* lines will indicate which, if any, of the keys is operated.

Standard input and output interface circuits are available for most microprocessors and we shall assume that one is used here, so arranged that an eight-bit input port appears mapped into memory space at location 8000 hexadecimal and an eight-bit output port at 8001. It is sensible to choose orderly connections between the keypad and the ports. Probably the best arrangement here is to use the least significant four-bits of the output port for the *Y* lines, so that the bit positions in both input and output words can correspond directly with the *X* and *Y* coordinates. Software debouncing of the keys is appropriate here because there will be adequate time, the microprocessor already exists, and a single debounce routine can serve for all keys. Hardware debouncing would require many circuits and would be difficult anyway with this contact scheme.

An outline flow diagram for the program is shown in figure 8.7. At this stage the objective is to represent the overall structure without becoming too concerned with detail. It will be immediately obvious that we must specify what happens if the program finds a key already pressed when it first scans the input port. In most cases it will be safer to wait until the keypad is inactive before starting the interrogation and so the flow diagram shows an initial test for activity. If the keypad is active the program loops back to the beginning and tries again. After the keypad becomes free, a new loop commences, waiting for the next key press. When this is detected, the program stores the key value, waits for a few milliseconds and tests the keypad again. If the key value is still the same, it is assumed to be valid and the program returns.

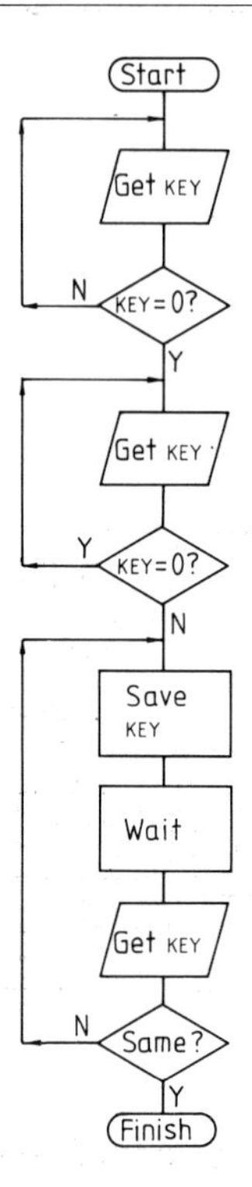

Figure 8.7 Flow diagram for the key pad interrogation.

Now some of the detail can be filled in. It is clear that all three wait loops scan the keypad. The first is not concerned with which particular key is pressed, and here it is tempting to write a simple test just to determine activity. However, we shall have to write a more complicated routine to evaluate the key position in the other loops, and this test can effectively include the first. Writing a single subroutine to serve for all three will save time and memory space, and the program will be more reliable.

There are four *Y* lines which must be pulsed in turn. During each pulse the *X* lines must be examined. This suggests that a loop controlled by a counter is a good way of implementing the test. The key value routine in figure 8.8 shows how this might be achieved by writing a pattern containing a single 1 to the output port. First, Y_0 is taken to 1 while the other *Y* lines remain at 0. The program now tests the *X* lines to determine if any of the bottom row of contacts was made. If all the *X* lines are zero, the pattern is shifted left and a counter YPOS incremented. Y_1 is now raised to 1 and the next key row tested. The loop continues until all the rows have been tested or until a contact is detected.

This loop terminates if an active *X* line is found, and the key position is evaluated by another shifting operation within a loop. The eight-bit code from the *X* lines is shifted right and the counter XPOS is incremented each

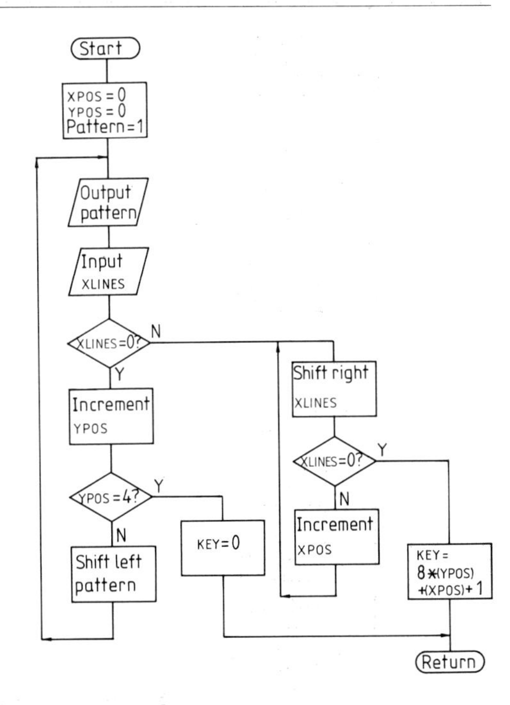

Figure 8.8 Flow diagram for the key position evaluation.

time the loop is traversed until the active bit is shifted out of the accumulator.

XPOS and YPOS now contain the coordinates of the key and a value can then be calculated from (XPOS) + 8*(YPOS) + 1 where (XPOS) indicates the contents of counter XPOS. No explicit multiplication instruction will be needed as three left shifts of a binary word are equivalent to multiplication by eight. If no active X line is found a value of zero is returned to indicate that the keypad is idle.

A complete program is given in figure 8.9 using mnemonics for the instructions and labels for memory addresses. Symbolic names defined at the beginning of the program are used for the addresses in memory where variables are stored to increase readability. All microprocessor manufacturers devise a set of mnemonics to describe the instructions their product can execute. We use here mnemonics which are common to several microprocessor instruction sets. In most cases they are self-explanatory. For example, ADD means add an operand to the value in the accumulator and leave the result in the accumulator; BSR means branch to the named subroutine; SHL is an instruction to shift a word one place left; EQU is a pseudo-instruction defining the value of a symbol. Immediate operands, indicated by the # symbol, or operand addresses follow the mnemonics. Comments, which should always be included in any program, describe what the program is doing.

```
XLINES     EQU $8000      /define the addresses at which time the X
YLINES     EQU $8001    • /and Y lines are mapped
XPOS       EQU 0          /assign location 0 to the variable XPOS
YPOS       EQU 1          /and location 1 for YPOS
PATTERN    EQU 2          /is in location 2
KEY        EQU 3          /is in location 3
/* Keypad interrogation routine */
/* waits for an idle keypad (KEY = 0) */
/* then waits for an active pad (KEY = <> 0) */
NOKEY      BSR PAD        /branch to the pad subroutine
           BNE NOKEY      /try again if not equal to zero
ISKEY      BSR PAD        /get the key value again
           BEQ ISKEY      /try again if idle
           RET            /return with the value
PAD        BSR VALUE      /interrogate the pad
NEXT       STA KEY        /save the key value
           LDX 255        /load a count in the index register
LOOP       DEC X          /decrement the count wasting time
           BNE LOOP       /loop back if not zero
           BSR VALUE      /get the key again
           CMP KEY        /is it the same?
           BNE NEXT       /try again if not
           RET            /return if it is
/* Key evaluation routine */
VALUE      CLR XPOS       /set XPOS to zero
           CLR YPOS       /and YPOS
           LDA #1         /load accumulator immediately with 1
           STA PATTERN    /store it in PATTERN
PULSE      LDA PATTERN    /get the pattern
           STA YLINE      /put it onto the Y lines
           LDA XLINES     /get the data on the X lines
           BNE XVALUE     /branch to the evaluation routine if >0
           INC YPOS       /increment YPOS
           LDA YPOS       /get YPOS and
           CMP 4          /compare with 4
           BEQ NOTKEY     /to check for the last row
           SHL PATTERN    /shift the pattern left 1 place
           BRA PULSE      /and pulse the next row
NOTKEY     CLRA           /clear accumulator if no key found
           BRA END        /and skip to the end
/* Key evaluation routine */
XVALUE     SHRA           /shift the accumulator right
           BNE XVALUE     /round again if the bit is still there
           LDA YPOS       /get the Y coordinate
           SHLA           /shift the accumulator
           SHLA           /left three places
           SHLA           /is times 8
           ADD XPOS       /add in the X coordinate
           ADD #1         /add 1 to adjust because 0,0 is key 1
END        RET            /and return
```

Figure 8.9 Assembler program to drive the keypad.

Assembler programming

This mnemonic form of the program can be translated into machine code by hand, but this tedious task is usually done with an assembler program. An assembler generates the operation codes for the instructions and assigns the defined values to addresses. It also calculates the proper addresses for the labels.

Nearly all microprocessors have assemblers provided by the manufacturers or obtainable from other sources. Assemblers can run either on the microprocessor for which the program is being developed or on some other computer. In the latter case they are known as cross assemblers.

Programs written in low level (assembler) language are directly concerned with the details of the microprocessor structure, its instruction set and the way in which it is connected to peripheral devices. Very often the purpose of a program becomes obscured by the fine detail of the implementation. Also, although there will be some similarities between assembler language programs written to perform the same task, with different microprocessors there will be many more differences. Therefore it is difficult to take a program written for one microprocessor and adapt it for use on another.

High level languages

High level languages aim for machine independence so that the programmer can concentrate on the algorithm required to solve the problem without worrying about microprocessor details. Programs should then become easier to write, more reliable and also portable from one machine to another. However, it has to be admitted that this ideal is not yet fully realised. Some adjustment of a program written in a high level language is usually needed before it can be run on a different microprocessor.

FORTRAN was one of the early high level languages, developed for scientific computation on large computers. It has been provided for some microprocessors, but it is not suited to systems with relatively small amounts of memory or for programs which need to process signals or control instrumentation. More recently BASIC, developed originally as a teaching language, has become popular and is available for the great majority of microprocessors.

The reason for the success of BASIC is not that it is a particularly good language, indeed it has many serious deficiencies, but that it can be implemented as a self-contained system in very small amounts of memory. Ten kilobytes is enough for an adequate BASIC with floating point arithmetic. A version with integer arithmetic only is possible in four kilobytes and restricted forms can be squeezed into as little as two kilobytes. This is small enough to integrate on the same chip as a microprocessor with some input and output circuitry, thus microcomputers are now available with a simple version of BASIC programmed into the on-chip ROM.

Programs written in a high level language have to be translated into machine code before they can be run on a microprocessor. Languages like FORTRAN are normally translated by a large compiler program, producing a block of self-contained machine code. Subsequently this block of code can be executed just like that produced from an assembler. Compilers vary widely in their efficiency. Most of those available for microprocessors generate far more code than would be written by a skilled programmer working at assembler level. The resultant program executes correspondingly slower, a factor between two and ten is common. Any alteration to a compiled program requires that the original text is edited and then recompiled to produce a new machine code module before the result of the change can be tested. Compilation is usually very slow, and the elimination of bugs is correspondingly tedious.

BASIC is usually an interpreter; that is, programs are held in memory as text, which is only translated into machine code while the program is being run. Interpretation has the advantage that changes to the program can be made and tested very rapidly. The disadvantage is that the text has to be translated every time the program is run, imposing a further speed penalty. Typically an interpreted program runs about ten times slower than a compiled version. So it might be as much as a hundred times slower than a program written at assembler level.

BASIC and FORTRAN were intended for computation rather than instrumentation and control. Both languages are cumbersome for applications where either manipulation of data at the bit and word level is required or where much interfacing to transducers and other peripherals is necessary. Also, neither encourages the production of reliable programs. PASCAL is a structured language, which leads to readable, reliable and maintainable programs. However, it too is intended for general purpose computation and does not fit well into the control and instrumentation field.

Other languages include CORAL and C which were originally developed for minicomputers. Both are good for control and are available for some of the more powerful microprocessors. They are well defined and standardised, meaning that the programs can be portable between different microprocessors. FORTH, which can be an assembler, interpreter or a compiler language is excellent for control applications, but not well enough standardised for programs to be easily portable.

The speed and program size penalties of compilers and interpreters ensure that some microprocessor programs are still written in assembler code. However, the ease of programming at high level, and the subsequent dramatic increase in clarity and reliability are a powerful incentive, and high level languages should now be used whenever possible.

9

Electronic Instrumentation

Instruments are essential in electronics to measure basic signal and component properties such as voltage, current, resistance, power, time and frequency. They may also be used indirectly to display other physical quantities such as flow rate, speed, mass and pressure when electronic methods of measurement are employed. Some instruments, such as the oscilloscope, are intended for the observation of signals rather than for accurate measurement. Yet the calibration of modern oscilloscopes can be better than 3% for both amplitude and time scales, hence they may be used in a limited way as measuring instruments. However, most measurements are made with instruments which specifically provide readings in either analogue or digital form.

9.1 Analogue and Digital Measuring Instruments

Although instruments may present their readings in analogue or digital form, the trend is towards digital display. Light emitting diode (LED), vacuum fluorescent (VF), and liquid crystal displays (LCD) using seven-segment digits are commonly used to provide numeric and very limited alpha character information. Dot matrix or 14- and 16-segment (starburst) devices are available for applications requiring full alphanumeric display. Occasionally the flexibility of a cathode ray tube (CRT) display may be useful, particularly if graphics are needed.

Analogue displays, where the reading is indicated by a pointer moving over a graduated scale, are still used in some electronic instruments. Although they are difficult to read accurately it is easy to judge the approximate value and to estimate trends. These are often preferred when a value has to be compared with a fixed reference, for example, when setting a voltage level to a particular value. Digital displays are easy to read when the reading is constant. However, if the value continually changes, or deviation from a fixed reference is required, or a trend needs to be recognised, these displays are less convenient. Nevertheless, these shortcomings are outweighed by the ruggedness and flexibility of digital displays. Moreover, digital displays are cheaper than analogue ones where high accuracy and resolution are necessary.

Accuracy and resolution

The accuracy of an instrument depends on both the precision of the references against which the measurement is made and the calibration and stability of any amplifiers or adjustable components in the measurement chain. Resolution is the smallest change in reading which can be observed, and must always be better than the accuracy.

Most analogue instruments are limited in both accuracy and resolution by the display which can rarely be read to better than about 0.1%; 1% is normal. Digital instruments can sometimes give a misleading impression of high accuracy because they are often provided with resolutions which are much better than the precision of the references or the overall calibration. Even so, most digital instruments do have higher resolution and accuracy than their analogue counterparts. For example, voltmeters are readily available which can measure to better than 0.01%.

Bar displays using a linear array of LED, VF or LCD segments are a compromise between pure analogue and pure digital indication. They are readily interfaced to digital circuitry, but have the advantage of analogue displays in showing trends and deviations. At present the resolution of bar displays is limited so they are not suitable for high precision indication. The provision of both a pure digital (numeric) display with an associated bar display is a compromise being increasingly adopted by instrument makers in an attempt to combine the best features of analogue and digital displays.

Voltmeters and multimeters

Voltage is the most common quantity measured in electronics. Low level signals will need amplification before they can drive a display, and sensitive voltmeters include precision amplifiers and calibrated potential dividers to provide multiple ranges. Auto-ranging, where the switching between ranges is achieved automatically, is a convenience feature found on the more expensive instruments.

Steady state or slowly varying quantities can be measured and displayed directly. High frequency signals are rectified and low pass filtered to produce a steady voltage proportional to the peak or RMS value of the waveform before measurement. Multimeters are voltmeters which include circuitry to convert current and resistance to voltage through Ohm's law, thereby combining several of the most useful measurement facilities in a single instrument.

Frequency and time measurement

Frequency can be measured very accurately using digital counting methods. The input signal is converted to a pulse train by amplification and limiting or by detecting zero crossings with a comparator. The stream of pulses is then gated into a BCD counter for a time determined from the divisions of a precision reference clock, as shown in figure 9.1. If the gate is open for a time t, the number of counts accumulated $N = tf$.

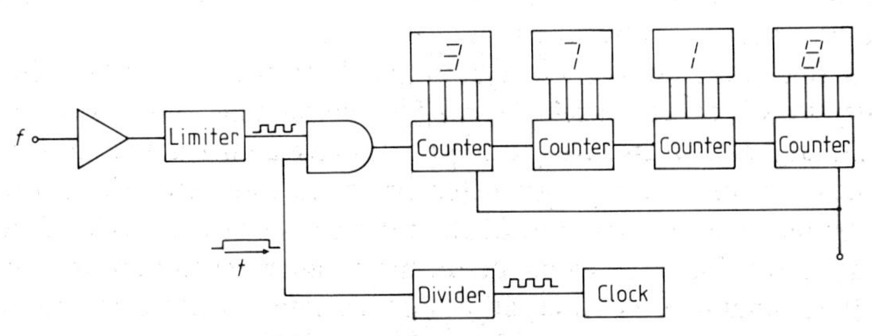

Figure 9.1 Four-digit frequency meter

Range changing can be achieved by selecting different division ratios and hence different gate times. Times are usually chosen to be decimal multiples or sub-multiples of one second to give direct scaling in hertz, kilohertz etc. The accuracy depends on the precision of the master clock which is determined by a quartz crystal. Absolute accuracy and stability of one part in a million is easily achieved.

A minor rearrangement of the frequency meter circuit can be used for time measurement. The shaped input signal gates a stream of reference clock pulses into the counter. Again, suitable division of the master clock frequency will give counter readings directly scaled in microseconds, milliseconds etc.

9.2 Digital to Analogue Conversion

Most electronic quantities are analogue and need conversion before they can be processed or displayed in digital form. It is convenient to consider first the reverse process of digital to analogue conversion which is often required in its own right, and may also itself be a part of the analogue to digital conversion process.

Conversion of a digital word to an analogue voltage can be accomplished with a simple extension of the operational amplifier summation circuit. Figure 9.2 shows a digital to analogue converter (DAC) which can accept a four-bit pure binary word *abcd* whose value is N. The input resistors have values of R, $2R$, $4R$ and $8R$ to provide input currents corresponding to the weights of the digits in the word. If the logic levels of the word are 0 and H for logic 0 and 1 respectively, then the current I_a that flows as a result of the most significant bit of the word is

$$I_a = D_a H/R.$$

The current I_b corresponding to the next most significant bit is

$$I_b = D_b H/2R$$

where D is the value of the digit, either 0 or 1. Similar expressions hold for I_c and I_d. The total current I is the sum of I_a to I_d, and so

$$V_o = (-HR_f/R)N. \tag{9.1}$$

Proper operation of this circuit depends on good definition of the levels 0 and H which are used as references in the derivation of the output signal. Logic levels obtained directly from gates are not accurate enough and it will normally be necessary to buffer the inputs with BPT or MOST switches and provide a stable voltage reference.

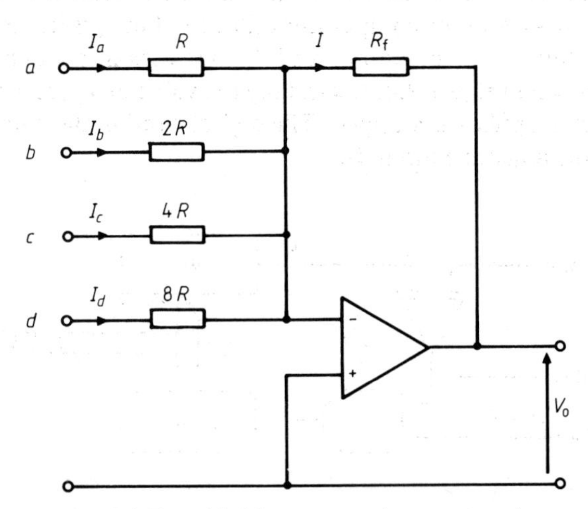

Figure 9.2 Four-bit binary to analogue converter.

This four-bit circuit can be extended to longer wordlengths but difficulties arise in achieving accurate matching of the input resistors over a wide range. Circuits using ladder networks which require only one or two matched values of resistance are preferred when conversion of six or more bits is required.

Many applications are satisfied by DAC with wordlengths in the six-bit to ten-bit range, corresponding to resolutions of 3% to 0.1%. 14 or 16 bits may be needed for very high accuracy conversion.

The output of a converter takes time to respond to an input word change. Operating speeds usually range from about 100 ns to 10 μs, although specialised converters operating down to about 1 ns are available with restricted wordlengths. The cost of converters rises rapidly with increasing speed and longer wordlength.

Complete DAC are made in integrated circuit form and may include voltage references, buffer amplifiers for the output signal, and input registers to allow easy interfacing with microprocessor circuits. Converters which operate on BCD rather than pure binary input signals are also available and are often more convenient in instrumentation applications.

9.3 Analogue to Digital Conversion

A common method of making an analogue to digital converter (ADC) uses a DAC in a feedback configuration. Figure 9.3 shows a simple arrangement in which the output from a counter is converted to a voltage and compared with the input signal. The output from the comparator controls the clock pulse stream into the counter. A conversion is initiated by resetting the counter to zero. V_c becomes 0 and the comparator indicates that V_c is less than V_i. Clock pulses then flow into the counter and V_c increases in a ramp or staircase fashion as shown in figure 9.4. Eventually the value of V_c just exceeds V_i and the flow of clock pulses is stopped. The digital word in the counter is now a measure of the analogue input V_i.

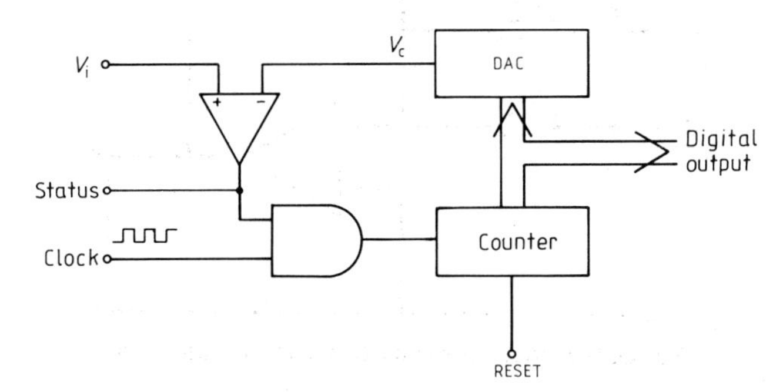

Figure 9.3 A staircase analogue to digital converter.

The comparator output S can be used to signal the status of the converter. Immediately after the conversion is initiated S goes high to indicate that the

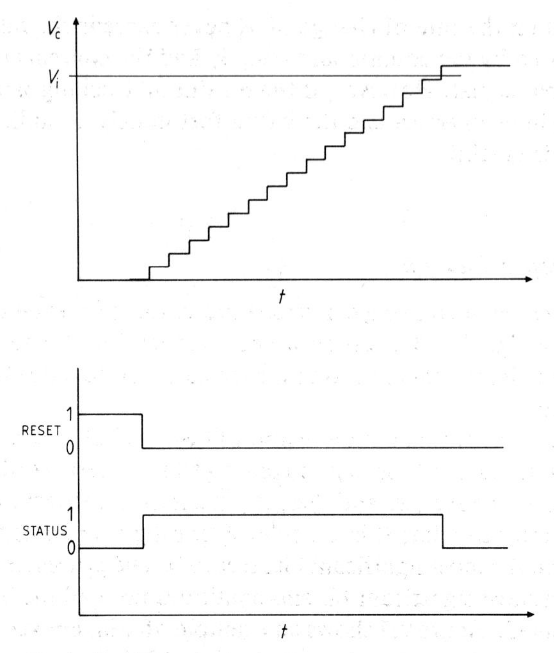

Figure 9.4 Waveforms in the staircase converter.

converter is busy. S returns to the low state when conversion is complete and the digital output word is valid.

This ramp converter is rather slow because the counter is reset for every conversion. If V_i does not change much between conversions, a faster measurement can be achieved by a simple modification of the basic circuit. Figure 9.5 shows an arrangement where the comparator output controls an up/down counter which is continuously clocked. If V_c is below V_i the counter is forced to count up and if V_c is above V_i the counter counts down.

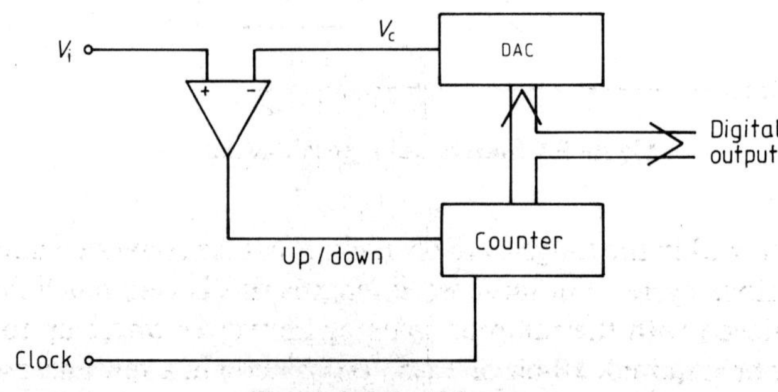

Figure 9.5 Tracking ADC.

Provided that the rate of change of V_i never exceeds the rate at which V_c can be adjusted by the counter and DAC, V_c and the contents of the counter track the input signal. However, if the maximum tracking rate is exceeded, the output will be in error, and there is unfortunately no indication that the output here is invalid.

Successive approximation

Neither staircase nor tracking converters are suitable for rapidly changing or discontinuous signals. They are slow because they alter the test digital word from the least significant end, so that each clock pulse only changes V_c by a small amount.

The successive approximation method illustrated in figure 9.6 is quicker because it starts at the most significant end of the test word. The control logic first clears the register and then sets the most significant bit to 1. If the comparator indicates that V_c is less than V_i, the digit is retained. Otherwise it is rejected and the most significant bit reset to 0. The process is then repeated with the next most significant bit and continues until all the bit values have been determined. Figure 9.7 shows an example of 4-bit conversion where the input corresponds to a value just greater than 1001 (9_{10}).

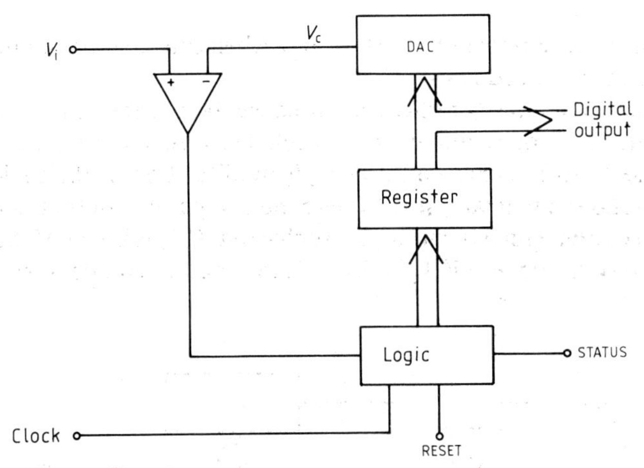

Figure 9.6 Successive approximation ADC.

Any input within the range of the converter can be converted into an n-bit word in n clock cycles. For large word lengths this is very much faster than can be achieved with the ramp or tracking converter where up to 2^n clock pulses may be required. 10-bit or 12-bit conversion in a few microseconds is readily attained.

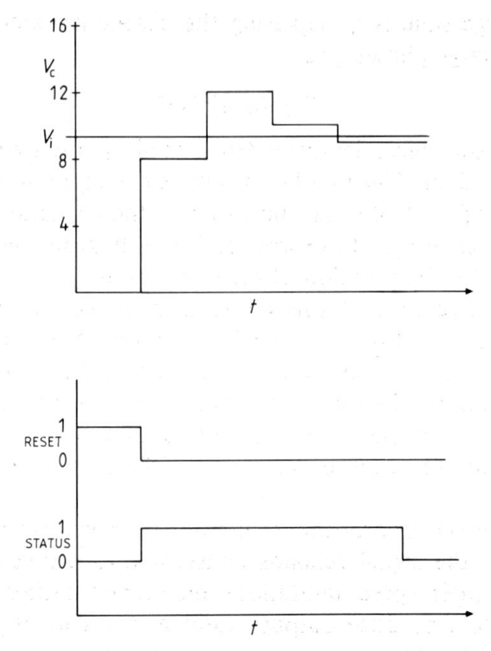

Figure 9.7 Waveforms in the successive approximation ADC.

Integrating ADC

Many digital meters needing high accuracy conversion but requiring only a slow conversion rate of perhaps a few times a second use the charging of a capacitor as the basis of analogue to digital conversion. In one common form the input voltage is converted to a current which charges a capacitor at a constant rate for a fixed time t_c. The capacitor is then discharged back to its original state with a current derived from a reference voltage V_r as shown in figure 9.8.

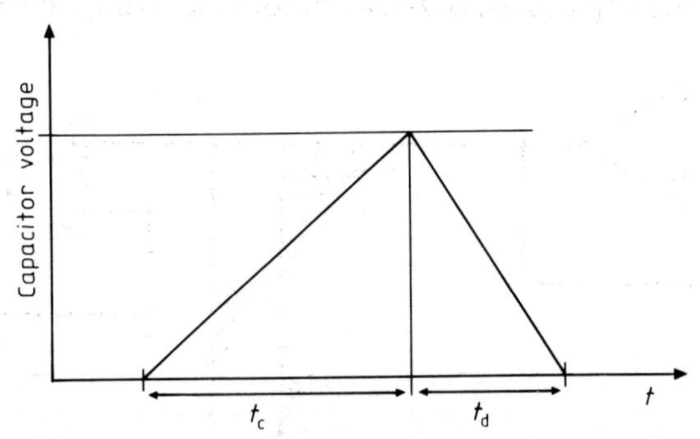

Figure 9.8 Dual-slope integrating ADC illustrating charge and discharge times.

If the discharge time is t_d, equating the charge transferred during the charge and discharge phases gives

$$V_i = (t_d/t_c)V_r. \tag{9.2}$$

Both t_d and t_c can conveniently be determined by counting pulses derived from a reference clock. The number of pulses corresponding to t_d is then a digital measure of V_i. Note that the value of the capacitor and the clock frequency do not enter into the expression for V_i. It is only required that they remain constant for the duration of the conversion.

A particular advantage of this conversion method is that the output depends on the average (integrated) value of V_i over the conversion process rather than the instantaneous value at some particular time. The effect of high frequency noise is then substantially reduced because it approximately averages to zero over a long period. The effect of hum is also reduced if the charge time is made a multiple of the period of the mains supply (20 ms in the UK).

All of these converters produce results which are valid to within one least significant bit if the signal remains constant throughout the conversion period. Within their speed limitations mentioned earlier, staircase and tracking converters produce outputs valid at the end of the conversion period even if the signal is not constant. However, the successive approximation converter determines the individual bits separately at different times and may produce serious errors if the input signal changes substantially.

Sample and hold circuits

If the input signal is changing rapidly during the conversion period it may be desirable to sample the input signal over a short time and hold that value constant during the conversion process. Such sample and hold circuits use charge storage on a capacitor and are readily available in IC form (see figure 9.9). During the sample phase S/H (sample or hold) is high and the capacitor

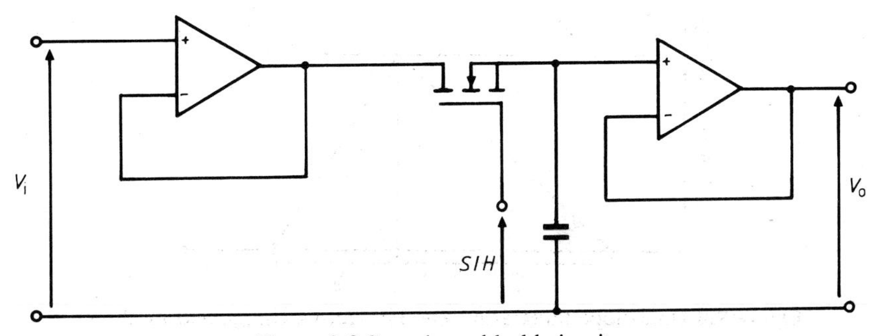

Figure 9.9 Sample and hold circuit.

is connected to the input signal through the MOST switch. A unity gain voltage follower is included to avoid loading effects. In the hold phase, S/H goes low and the capacitor is isolated from the input signal and the stored value is available at the output through another voltage follower.

DAC and sample and hold circuits are readily available in IC form in a variety of speed and accuracy combinations. Some are also made which include LED, VF or LCD decoding and driving circuitry. Thus it is possible to make a complete digital voltmeter with just one chip, a display and a few external components.

9.4 Data Acquisition and Logging

There are many simple instrumentation applications where it is sufficient for the readings to be presented in digital or analogue form directly to the user. However, in more complicated applications where the volume of data is large, or readings need to be taken rapidly, some automation of the process is highly desirable.

It may also be necessary to process the data before it can be used. For example, the continuous monitoring of the temperature and pressure in an industrial chemical plant might yield tens of thousands of readings, most of which are normal. Only occasionally need an unusual value be brought to the attention of a human operator.

The trend towards digital rather than analogue techniques has made it easier to manipulate data obtained from instruments. If a reading is in digital form it is nearly as easy to transmit, process or store as it is to display directly. Data loggers which can collect large amounts of information from many sources, often remote from the logger, are widely used in laboratory and industrial instrumentation. They may accept analogue or digital signals, though in the first case they will be conditioned and converted to digital data. A microcomputer is often incorporated to process the results.

Most data loggers have a real time internal clock providing timing signals so that they can scan the signal inputs and take readings at regular intervals. Small amounts of data can be stored in non-volatile semiconductor memory such as EPROM or battery-backed CMOS RAM. Larger volumes may be put onto bulk storage such as magnetic disc or tape.

9.5 Computer Control of Instrumentation

Many instruments now have provision for control of their functions by external digital signals. A common example might be a digital multimeter in which the function—voltage, current, resistance etc—and the range can be programmed by suitable commands. Also, it may be possible to instruct the meter to take a measurement on receipt of an initiating signal, or perhaps to take a series of readings at a predetermined rate. Other laboratory equipment such as power supplies, signal sources, plotters and chart recorders may also be provided with means for external control.

Computer control is obviously valuable when a number of instruments have to be controlled and their readings collected or when measurements must be taken at high rates. The computational facility may also be exploited for equation solving, result processing, drift or offset correction or linearisation or making intelligent decisions such as how many readings to take to achieve a desired degree of accuracy. Even for less demanding tasks computer control is so convenient that it is becoming the normal way of operating instrumentation.

9.6 Standard Interfaces and Buses

In order to allow interchangability between instruments which may need to be interconnected, it is sensible to adopt some form of standard interface. The standard may relate to the physical details of the interface such as the number and function of the signal lines, the voltage levels and timing of the signals, or it may define the form of the data and control information which can be transmitted, or both.

Some digital instruments are provided with a BCD output using TTL compatible signal levels. The output contains the displayed reading possibly with range information. Sometimes, control lines for reset and reading initiation are provided. It is then relatively straightforward to connect the instrument to a computer, control the operation and collect the readings. This may be adequate if only one or two instruments are involved, but not if a number are used in an extended arrangement.

IEE-488 interface bus

One popular standard interface allowing control both of the instruments and exchange of data with the control computer is the IEE-488 bus. This was originally developed by Hewlett Packard for their own range of instruments but has now been widely adopted by other manufacturers. It is sometimes also known as the general purpose interface bus (GPIB).

The specification of the IEE-488 bus defines the form of the physical interconnection between the instruments, the signal levels and the protocol which governs the exchange of information over the bus. This guarantees that any number of instruments up to the maximum of 15 can be interconnected and interchanged without regard to their precise function, speed of response or make. One of the instruments is a controller, often a microcomputer, which organises the operation of the other instruments and collects the results.

The bus uses eight data lines and eight control lines. The control lines carry signals which regulate the flow of information on the data lines allowing for variable response times from different instruments and resolving contention problems when more than one instrument wishes to use the bus at the same time. Eight lines are needed to allow for a wide variety of control situations but many instruments use only a subset of four of five or them.

A particular feature of the IEE-488 interface which ensures reliable communication between instruments is the 'handshake' in which the transmitter signals its intention to send a message, the receiver responds by signalling back its ability to accept, the message is sent and finally the receiver acknowledges receipt of the message.

Control data and readings are transmitted as a stream of eight-bit words on the parallel data bus. The precise form of the data is not specified, but many instruments use ASCII text characters. So, for example, a multimeter might return a reading of 13.45 mV as the character stream '+13.45E−3'. In most cases it would be unnecessary to transmit the instrument function—voltage measurement in this case—as well, because the multimeter would have been programmed previously by the controller.

In nearly all applications the detailed operation of the information exchange is handled by hardware and software built into each instrument and need not concern the user. Low level language subroutines or high level language commands are usually provided for the controller to send streams of function control signals and to accept measurement readings. Programs can then readily be written to drive a particular experimental arrangement.

Interpreted BASIC is commonly provided as a high level language for control and result processing. It is adequate for the majority of instrumentation applications but may be too slow when readings are required at a rate of more than about 10–20 s^{-1}. In these cases a compiled language, or low level programming must be used.

Index